BestMasters

Mit „**BestMasters**" zeichnet Springer die besten Masterarbeiten aus, die an renommierten Hochschulen in Deutschland, Österreich und der Schweiz entstanden sind. Die mit Höchstnote ausgezeichneten Arbeiten wurden durch Gutachter zur Veröffentlichung empfohlen und behandeln aktuelle Themen aus unterschiedlichen Fachgebieten der Naturwissenschaften, Psychologie, Sozialwissenschaften, Technik und Wirtschaftswissenschaften. Die Reihe wendet sich an Praktiker und Wissenschaftler gleichermaßen und soll insbesondere auch Nachwuchswissenschaftlern Orientierung geben.

Springer awards **"BestMasters"** to the best master's theses which have been completed at renowned Universities in Germany, Austria, and Switzerland. The studies received highest marks and were recommended for publication by supervisors. They address current issues from various fields of research in natural sciences, psychology, social sciences, technology, and economics. The series addresses practitioners as well as scientists and, in particular, offers guidance for early stage researchers.

Julien Spahn

Pioneering Raman Transition Techniques in Collinear Laser Spectroscopy for High-Precision Experiments

Julien Spahn (ORCID)
Institute for Nuclear Physics
Technische Universität Darmstadt
Darmstadt, Germany

ISSN 2625-3577   ISSN 2625-3615   (electronic)
BestMasters
ISBN 978-3-658-50604-9   ISBN 978-3-658-50605-6   (eBook)
https://doi.org/10.1007/978-3-658-50605-6

This Springer Spektrum imprint is published by the registered company Springer Fachmedien
Wiesbaden GmbH, part of Springer Nature.
The registered company address is: Abraham-Lincoln-Str. 46, 65189 Wiesbaden, Germany

If disposing of this product, please recycle the paper.

Acknowledgments I will try to keep at least this part short and hope that I do not forget someone:

I would like to thank **Prof. Wilfried Nörtershäuser** and the entire **Laser-SpHERe group** for giving me the opportunity to write my master's thesis in this group. Thanks for the lively discussions, the continuous support and always taking the time to discuss my questions.

Special thanks to **Hendrik Bodnar** and **Kristian König** for helping me during the measurements and proofreading this thesis.

I also thank the **group of Ruben de Groote from KU Leuven** for providing the diode laser, without which many measurements in this work would not have been possible. Special thanks to **Phillip Imgram** for driving the laser all the way to Darmstadt and back, and to **Robbe Van Duyse** for taking a week of his time to help me in the preparation of the measurements in the double Raman scheme.

Last but not least I would like to say that I have learned a lot over the last year. It has been a lot of fun, and I look forward to learning even more and tackling future challenges.

Competing Interests The author has no competing interests to declare that are relevant to the content of this manuscript.

Contents

List of Figures

Motivation **1**

Collinear laser spectroscopy is a powerful method that is being used since the 1970s [1] to investigate optical transitions all across the nuclear landscape. It is especially well suited to investigate short-lived radioactive isotopes [2, 3]. The precise measurement of atomic transition frequencies gives access to a variety of nuclear properties, such as the charge radius via the isotope shift or the nuclear spin, as well as the magnetic dipole and the electric quadrupole moment through the hyperfine structure [4–6]. Those can be used to investigate the nuclear structure, *e.g.*, shell closures in magic nuclei [7, 8], and to benchmark nuclear theory calculations.

Raman transitions (inelastic two-photon scattering) were discovered over a century ago [9, 10] and various applications utilizing them have emerged over time. Those include solid-state spectroscopy [11], microscopy on the nanoscale [12], and the cooling of ultra-cold atoms [13].

Recent atomic calculations in Ca^+, performed by Neumann et al. [14], indicate that the use of Raman transitions could be extended to collinear laser spectroscopy. Since the width of Raman transitions is orders of magnitude smaller than the natural linewidth of an allowed dipole transition, this could increase the accuracy of spectroscopic measurements. In particular, Neumann et al. proposed the use of Raman transitions in a "Raman velocity filter" to selectively prepare the population of a meta-stable state for subsequent high precision-spectroscopy. This opens the possibility of improving high-voltage measurements performed at the COALA beamline (<u>Co</u>llinear <u>A</u>pparatus for <u>L</u>aser Spectroscopy and <u>A</u>pplied Science) at TU Darmstadt [15]. So far, the only application of collinear Raman spectroscopy was realized using a single laser and its rf-shifted sideband to transfer the population between hyperfine states of the same fine structure level [16]. This enabled to determine small magnetic dipole and electric quadrupole moments that could not be resolved by conventional collinear laser spectroscopy. It has not yet been demonstrated that

J. Spahn, *Pioneering Raman Transition Techniques in Collinear Laser Spectroscopy for High-Precision Experiments*, BestMasters,
https://doi.org/10.1007/978-3-658-50605-6_1

Raman transitions can be driven in an ion beam using two separate lasers, which would expand this method to a wide range of transitions. This would enable the investigation of elements, that do not offer allowed dipole transitions in the optical regime, and hence allow investigating their nuclear properties. However, previous attempts at the COALA setup at TU Darmstadt to do so have been unsuccessful [17].

This work provides new calculations and simulations on driving Raman transitions with two different lasers at a collinear setup. Simulations will include ion energy distributions and spatial intensity distributions. First experimental results in the $S_{1/2}$-$P_{3/2}$-$D_{3/2}$ and $S_{1/2}$-$P_{3/2}$-$D_{5/2}$ Λ-schemes of $^{88}Sr^+$ are presented. A newly developed Doppler-free double Raman scheme is used to measure the $D_{3/2} \rightarrow D_{5/2}$ transition. The measurements performed in this work investigate the systematic uncertainties of this method as well as its experimental limitations and will demonstrate the feasibility and the advances of this approach.

Theoretical Background 2

In this chapter, a brief introduction to collinear laser spectroscopy is given, and Raman transitions as two-photon Rabi-oscillations in a three-level system are derived and discussed.

Finally, an overview of the part of the Sr^+ level scheme relevant to the measurements done within this work is given.

2.1 Collinear Laser Spectroscopy

Laser spectroscopy is the laser-induced excitation of an atomic or molecular transition and its subsequent detection. In the simplest case, a photon from a laser is absorbed if its frequency matches the energy difference between two electronic states of an atom. The atom is then excited to the higher-lying state. This can be detected by measuring absorption spectra or by detecting the photons emitted when the atom spontaneously decays back into the lower-lying state.

The latter is known as fluorescence spectroscopy and the method of detection most commonly employed in collinear laser spectroscopy. The laser frequency is scanned across the resonance, and a resonance signal is detected in the form of a peak in the fluorescence light. More elaborate detection schemes with higher sensitivity rely on particle detection like the ROC method (radioactive detection of the optically pumped ions after state selective charge exchange) developed at COLLAPS at ISOLDE-CERN [18, 19] or collinear resonance ionization spectroscopy (CRIS) [20].

In collinear laser spectroscopy, laser spectroscopy is performed on a beam of ions or neutral atoms with a laser in parallel (collinear, col.) and/or anti-parallel (anticollinear, acol.) configuration. This enables spectroscopy on exotic radioactive

J. Spahn, *Pioneering Raman Transition Techniques in Collinear Laser Spectroscopy for High-Precision Experiments*, BestMasters, https://doi.org/10.1007/978-3-658-50605-6_2

isotopes with lifetimes of a few 10 ms that are too short-lived to be investigated in trap experiments. Compared to spectroscopy on trapped ions, this method has two main disadvantages: The beam cannot be cooled to sub-K temperatures, causing Doppler broadening, and the observed transition frequency is Doppler shifted.

Doppler broadening is the broadening of the linewidth due to the velocity width Δv induced by the energy width ΔE of the ion beam leaving the ion source [21]. The impact of Doppler broadening in collinear spectroscopy will be further discussed in Sect. 2.3.1.

Secondly, the transition frequency measured in the laboratory ω_{lab} is not the rest-frame transition frequency ω_0, but is Doppler shifted,

$$\omega_{\text{lab}} = \omega_0 \cdot \frac{1}{\gamma(1 - \beta \cos\alpha)}, \tag{2.1}$$

with the Lorentz factor $\gamma = \frac{1}{\sqrt{1-\beta^2}}$ and $\beta = \frac{v}{c}$ where v is the speed of the ions in the laboratory frame, c is the speed of light in vacuum and α is the angle between the ion and the laser beam.

In the case of collinear ($\alpha = 0$) or anticollinear ($\alpha = \pi$) geometry this simplifies to

$$\omega_{\text{col/acol}} = \omega_0 \cdot \gamma(1 \pm \beta). \tag{2.2}$$

Alternatively, for ions of mass m and charge q, accelerated electrostatically by a voltage U_{acc}, this can be rewritten as

$$\omega_{\text{col/acol}} = \omega_0 \cdot \left[1 + \frac{qU_{\text{acc}}}{mc^2} \left(1 \pm \sqrt{1 + 2\frac{mc^2}{qU_{\text{acc}}}} \right) \right]. \tag{2.3}$$

This voltage dependency of the measured transition frequency can also be used to measure high-voltages at ppm accuracy if the rest frame transition frequency is well known, as has been demonstrated at COALA within the ALIVE (Accurate Laser Involved Voltage Evaluation) experiment [15, 22, 23].

Extracting the rest frame transition frequency ω_0 from the measured transition frequency $\omega_{\text{col/acol}}$ and the acceleration voltage U_{acc} employing Eq. (2.3), is in practice limited in precision, due to systematic uncertainties of the experimental setup such as contact potentials and the starting velocity of the ions in the ion source.

Instead, the transition frequency is measured both in collinear and anticollinear geometry in fast alternation (quasi-simultaneous). Since $\gamma^{-2} = (1 + \beta)(1 - \beta)$,

Eq. (2.2) yields, that the rest frame transition frequency can be extracted from the product of the two resonant laser frequencies

$$\omega_{\text{col}} \cdot \omega_{\text{acol}} = \omega_0^2. \tag{2.4}$$

Scanning the frequency of a laser is slow, limits the laser stability, and is only possible over a limited range. Therefore, instead of scanning the laboratory-frame laser frequency, often the Doppler-shifted ion rest-frame laser frequency is scanned by scanning the ion energy. This is known as Doppler tuning.

2.2   Raman Transitions in a Three-Level System

A three-level system interacting with two lasers is shown in Fig. 2.1. The first laser (frequency ω_{L1}, wave vector $\vec{k}_{\text{L1}}$, phase ϕ_1) couples the ground state $|1\rangle$ (energy $\hbar\omega_1$) to the excited state $|1\rangle$ (energy $\hbar\omega_3$) via the dipole interaction and the second laser (frequency ω_{L2}, wave vector $\vec{k}_{\text{L2}}$, phase ϕ_2) couples the excited state $|3\rangle$ to a lower lying meta-stable state $|2\rangle$ (energy $\hbar\omega_2$).

A Raman transition is an inelastic two-photon scattering, which can be understood as the simultaneous stimulated absorption of a photon of frequency ω_{L1} and stimulated emission of a second photon ω_{L2}, resulting in a transition from some initial state $|1\rangle$ to a final state $|2\rangle$ through a virtual intermediate state, that is detuned from the real intermediate state $|3\rangle$ by a large amount Δ.

This is achieved by detuning both lasers from their single-photon resonance by $\Delta_{11} = \omega_{13} - \omega_{\text{L1}}$ and $\Delta_{22} = \omega_{23} - \omega_{\text{L2}}$, as shown in Fig. 2.1. $\omega_{nm} = \omega_m - \omega_n$ is the rest-frame transition frequency from level $|n\rangle$ to $|m\rangle$. If the two-photon resonance condition $\Delta_{11} = \Delta_{22}$ is fulfilled, this results in a two-photon Rabi oscillation between the states $|1\rangle$ and $|2\rangle$.

Let $|\Psi\rangle_{\text{RWA}}$ be an arbitrary state in the basis $\{|1, \vec{p}\rangle, |2, \vec{p}+\hbar(\vec{k}_{\text{L1}}-\vec{k}_{\text{L2}})\rangle, |3, \vec{p}+\hbar\vec{k}_{\text{L1}}\rangle\}$, given by

$$|\Psi\rangle_{\text{RWA}} = \begin{pmatrix} a_1 \\ a_2 \\ a_3 \end{pmatrix} = a_1|1, \vec{p}\rangle + a_2|2, \vec{p}+\hbar(\vec{k}_{\text{L1}}-\vec{k}_{\text{L2}})\rangle + a_3|3, \vec{p}+\hbar\vec{k}_{\text{L1}}\rangle, \tag{2.5}$$

where the basis has been expanded to include momentum space and $\vec{p}$ is the initial momentum of the atom in the ground state. An intuitive explanation for this can be given: When absorbing/emitting a photon, the atom also absorbs/loses its momentum

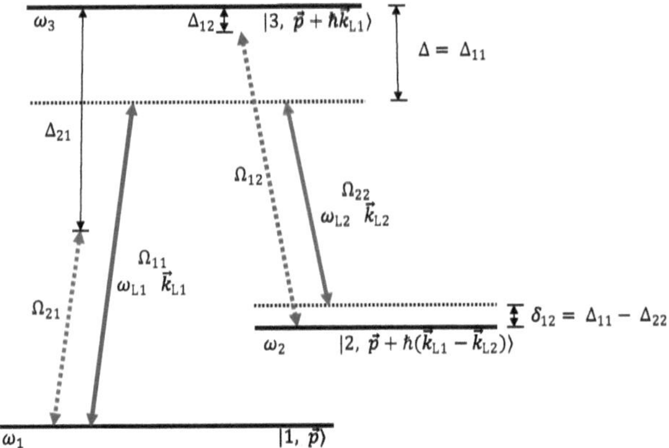

Fig. 2.1 Illustration of a Raman transition in a Λ-scheme. The arrows indicate the coupling of the laser $j \in \{1, 2\}$ (frequency ω_{Lj}, wavevector $\vec{k}_{Lj}$) with the transition between level $|n\rangle$, $n \in \{1, 2\}$ and $|3\rangle$ with the Rabi frequency Ω_{nj} and the single-photon detuning Δ_{nj}. The full arrows correspond to the couplings driving the Raman transition, while the dashed arrows correspond to the interactions that only result in additional AC-Stark shifts

due to momentum conservation. Hence, if an atom with momentum $\vec{p}$ absorbs a photon with momentum $\vec{k}_{L1}$, it then has the momentum $\vec{p} + \hbar\vec{k}_{L1}$. After consecutive emittance of a photon with momentum $\vec{k}_{L2}$ it has the momentum $\vec{p} + \hbar(\vec{k}_{L1} - \vec{k}_{L2})$. An analytic derivation of this is given in [24].

In the present application with large Δ, the population of $|3\rangle$ is assumed to be negligible, and spontaneous decay from the metastable state $|2\rangle$ is omitted.

It is assumed that the first laser only couples $|1, \vec{p}\rangle$ to $|3, \vec{p}+\hbar\vec{k}_{L1}\rangle$ and the second laser only couples $|3, \vec{p}+\hbar\vec{k}_{L1}\rangle$ to $|2, \vec{p}+\hbar(\vec{k}_{L1} - \vec{k}_{L2})\rangle$. The Hamiltonian of such a three-level system, using minimal coupling and applying the Power-Zienau-Woolley transformation as well as the rotating wave approximation, can be decomposed into a kinetic, an atomic, and an interaction part, and is given by

$$\hat{H}_{\text{RWA}} = \hat{H}_{\text{kin}} + \hat{H}_{\text{atm}} + \hat{H}_{\text{int}}. \tag{2.6}$$

For a classical electric field of the lasers of the form $\vec{E} = \sum_{j \in 1,2} \frac{1}{2}\vec{E}_i\, e^{i(\vec{k}_{Lj}\cdot\vec{r} - \omega_{Lj}t + \phi_j)} +$ c.c., the different contributions are given by [24]

$$\hat{H}_{\text{kin}} = \begin{pmatrix} \frac{\vec{p}^2}{2M} & 0 & 0 \\ 0 & \frac{(\vec{p}+\hbar(\vec{k}_{L1}-\vec{k}_{L2}))^2}{2M} & 0 \\ 0 & 0 & \frac{(\vec{p}+\hbar\vec{k}_{L1})^2}{2M} \end{pmatrix}, \tag{2.7a}$$

$$\hat{H}_{\text{atm}} = \begin{pmatrix} \hbar\omega_1 & 0 & 0 \\ 0 & \hbar\omega_2 & 0 \\ 0 & 0 & \hbar\omega_3 \end{pmatrix}, \tag{2.7b}$$

$$\hat{H}_{\text{int}} = \begin{pmatrix} 0 & 0 & \frac{\hbar\Omega_{11}}{2}e^{i(\omega_{L1}t-\phi_1)} \\ 0 & 0 & \frac{\hbar\Omega_{22}}{2}e^{i(\omega_{L2}t-\phi_2)} \\ \frac{\hbar\Omega_{11}^*}{2}e^{-i(\omega_{L1}t-\phi_1)} & \frac{\hbar\Omega_{22}^*}{2}e^{-i(\omega_{L2}t-\phi_2)} & 0 \end{pmatrix}. \tag{2.7c}$$

Where $\Omega_{nj} = -\frac{\langle n|\hat{\vec{d}}\cdot\vec{E}_j|3\rangle}{\hbar}$ is the Rabi frequency of the dipole coupling of level $|n\rangle$ to level $|3\rangle$, henceforth denoted as transition n, through the laser j, with the dipole operator $\hat{\vec{d}} = e\hat{\vec{r}}$. $\hat{\vec{r}}$ is the position of the electron relative to the nucleus.

For a Gaussian beam with total power P_j and beam waist w_j, the Rabi frequency is given by [25]

$$|\Omega_{nj}| = \frac{|d_{\text{eff},n}|}{\hbar}\sqrt{\frac{4P_j}{\pi\epsilon_0 c w_j^2}}, \tag{2.8}$$

where ϵ_0 is the vacuum permittivity, and $|d_{\text{eff},n}|$ is the effective dipole moment of the transition n. Assuming an isotropic laser field (equal components in all three possible polarizations) and neglecting hyperfine splitting, the effective dipole moment can be calculated through the reduced dipole matrix element $|\langle J|e\hat{\vec{r}}|J'\rangle|$ [25, 26]

$$|d_{\text{eff},n}|^2 = \frac{1}{3}|\langle J||e\hat{\vec{r}}||J'\rangle|^2, \tag{2.9}$$

where J and J' are the angular momentum quantum numbers of the initial and final state of the transition n. This reduced matrix element can be extracted from the relation

$$A_{J'J} = \frac{1}{\tau} = \frac{\omega_0^3}{3\pi\epsilon_0\hbar c^3}\frac{2J+1}{2J'+1}|\langle J|e\vec{r}|J'\rangle|^2, \tag{2.10}$$

using the Einstein coefficient $A_{J'J}$ of the transition or its lifetime τ as well as its transition frequency ω_0 [26].

A unitary transformation $|\Psi\rangle_{\mathrm{IF}} = \begin{pmatrix} b_1 \\ b_2 \\ b_3 \end{pmatrix} = \hat{U} |\Psi\rangle_{\mathrm{RWA}}$ is performed, in order to switch into the interaction frame of the atom. From the time-dependent Schrödinger equation, it follows that $\hat{H}_{\mathrm{IF}} = \hat{U} \hat{H}_{\mathrm{RWA}} \hat{U}^\dagger - i\hbar \hat{U} \frac{\partial}{\partial t} \hat{U}^\dagger$.

Here $\hat{U} = e^{i(\hat{H}_{\mathrm{kin}} + \hat{H}_{\mathrm{atm}})t/\hbar}$ is chosen. Since $[(\hat{H}_{\mathrm{kin}} + \hat{H}_{\mathrm{atm}}), \hat{U}] = 0$ and $i\hbar \hat{U} \frac{\partial}{\partial t} \hat{U}^\dagger = (\hat{H}_{\mathrm{kin}} + \hat{H}_{\mathrm{atm}})$,

$$
\hat{H}_{\mathrm{IF}} = \hat{U} \hat{H}_{\mathrm{int}} \hat{U}^\dagger = \begin{pmatrix} 0 & 0 & \frac{\hbar\Omega_{11}}{2} e^{-i(\Delta_{11}t+\phi_1)} \\ 0 & 0 & \frac{\hbar\Omega_{22}}{2} e^{-i(\Delta_{22}t+\phi_2)} \\ \frac{\hbar\Omega_{11}^*}{2} e^{i(\Delta_{11}t+\phi_1)} & \frac{\hbar\Omega_{22}^*}{2} e^{i(\Delta_{22}t+\phi_2)} & 0 \end{pmatrix}
$$
$$
= \begin{pmatrix} 0 & 0 & \frac{\hbar\Omega_{11}}{2} e^{-i(\Delta t+\phi_1)} \\ 0 & 0 & \frac{\hbar\Omega_{22}}{2} e^{-i((\Delta+\delta)t+\phi_2)} \\ \frac{\hbar\Omega_{11}^*}{2} e^{i(\Delta t+\phi_1)} & \frac{\hbar\Omega_{22}^*}{2} e^{i((\Delta+\delta)t+\phi_2)} & 0 \end{pmatrix},
$$
$$(2.11)$$

where the detunings from the single-photon resonances Δ_{11} and Δ_{22} are defined as

$$
\Delta_{11} = \omega_{13} - \left(\omega_{\mathrm{L1}} - \frac{\vec{p} \cdot \vec{k}_{\mathrm{L1}}}{M} - \frac{\hbar k_{\mathrm{L1}}^2}{2M}\right) =: \Delta \tag{2.12}
$$

$$
\Delta_{22} = \omega_{23} - \left(\omega_{\mathrm{L2}} - \frac{\vec{p} \cdot \vec{k}_{\mathrm{L2}}}{M} + \frac{\hbar k_{\mathrm{L2}}^2}{2M} - \frac{\hbar \vec{k}_{\mathrm{L1}} \cdot \vec{k}_{\mathrm{L2}}}{M}\right) \tag{2.13}
$$

and the detuning from the two-photon resonance $\delta = \Delta_{22} - \Delta_{11}$ is defined as

$$
\delta = (\omega_{23} - \omega_{13}) - \left(\omega_{\mathrm{L2}} - \omega_{\mathrm{L1}} - \frac{\vec{p} \cdot (\vec{k}_{\mathrm{L2}} - \vec{k}_{\mathrm{L1}})}{M} + \frac{\hbar}{2M}(\vec{k}_{\mathrm{L1}} - \vec{k}_{\mathrm{L2}})^2\right). \tag{2.14}
$$

Note that in equations (2.12) and (2.13) the term $\omega_{\mathrm{L1,2}} - \frac{\vec{p} \cdot \vec{k}_{\mathrm{L1,2}}}{M}$ is the classical Doppler shift from the laboratory laser frequency $\omega_{\mathrm{L1,2}}$ to the ion rest frame laser frequency. It has to be corrected to the relativistic Doppler effect according to equation (2.1). The terms quadratic in $k_{\mathrm{L1,2}}$ correspond to energy shifts induced by the absorption and loss of the photon momenta $\hbar k_{\mathrm{L1,2}}$.

2.2.1 From a Three- to a Two-Level Hamiltonian

By writing out the time-dependent Schrödinger equation $i\hbar\frac{\partial}{\partial t}|\Psi\rangle_{\text{IF}} = \hat{H}_{\text{IF}}|\Psi\rangle_{\text{IF}}$, one obtains a system of coupled differential equations describing the population dynamics of the three-level system:

$$\Leftrightarrow i\hbar\frac{\partial}{\partial t}b_1(t) = \frac{\hbar\Omega_{11}}{2}e^{-i(\Delta t+\phi_1)}b_3(t) \tag{2.15a}$$

$$i\hbar\frac{\partial}{\partial t}b_2(t) = \frac{\hbar\Omega_{22}}{2}e^{-i((\Delta+\delta)t+\phi_2)}b_3(t) \tag{2.15b}$$

$$i\hbar\frac{\partial}{\partial t}b_3(t) = \frac{\hbar\Omega_{11}^*}{2}e^{i(\Delta t+\phi_1)}b_1(t) + \frac{\hbar\Omega_{22}^*}{2}e^{i((\Delta+\delta)t+\phi_2)}b_2(t) \tag{2.15c}$$

For a large detuning $\Delta \gg \delta$, $\frac{\partial}{\partial t}b_1(t)$, $\frac{\partial}{\partial t}b_2(t) \sim \Omega_{ii}$, the Hamiltonian given in Eq. (2.11) can be reduced to an effective two-level system by integrating Eq. (2.15c), assuming $b_1(t)$, $b_2(t)$ to be constant. Solving for $b_3(t)$ and inserting the result into Eq. (2.15a) and (2.15b) yields

$$b_3(t) = -\frac{\Omega_{11}^*}{2\Delta}e^{i(\Delta t+\phi_1)}b_1(t)) - \frac{\Omega_{22}^*}{2(\Delta+\delta)}e^{i((\Delta+\delta)t+\phi_2)}b_2(t), \tag{2.16}$$

$$i\hbar\frac{\partial}{\partial t}b_1(t) = -\frac{\hbar|\Omega_{11}|^2}{4\Delta}b_1(t) - \frac{\hbar\Omega_{11}\Omega_{22}^*}{4(\Delta+\delta)}e^{i(\delta t-\phi_L)}b_2(t), \tag{2.17a}$$

$$i\hbar\frac{\partial}{\partial t}b_2(t) = -\frac{\hbar\Omega_{22}\Omega_{11}^*}{4\Delta}e^{-i(\delta t-\phi_L)}b_1(t) - \frac{\hbar|\Omega_{22}|^2}{4(\Delta+\delta)}b_2(t), \tag{2.17b}$$

where $\phi_L = \phi_1 - \phi_2$ is the relative phase of the two lasers. Assuming population of $|3, \vec{p} + \hbar\vec{k}_{\text{L1}}\rangle$ is negligible, since the detuning from both single-photon resonances is large, equation (2.17a) and (2.17b) can be rewritten into a 2x2 matrix equation to obtain an effective 2-level Hamiltonian:

$$\hat{H}_{\text{IF}} = (-\hbar)\begin{pmatrix} \Omega_1^{\text{AC}} & \frac{\Omega_{\text{R2}}}{2}e^{i(\delta t-\phi_L)} \\ \frac{\Omega_{\text{R1}}^*}{2}e^{-i(\delta t-\phi_L)} & \Omega_2^{\text{AC}} \end{pmatrix} \tag{2.18}$$

The diagonal entries of $\hat{H}_{\text{IF}}$ are the *AC*-Stark shifts Ω_n^{AC} of the state $|n\rangle$

$$\Omega_n^{AC} = \sum_{j=1,2} \frac{|\Omega_{nj}|^2}{4\Delta_{nj}}, \quad n \in \{1, 2\},$$ (2.19)

where the $n \neq j$ term results from the so far neglected coupling of the first laser to the $|2\rangle$ to $|3\rangle$ transition and of the second laser to the $|1\rangle$ to $|3\rangle$ transition (see Sect. 2.2.3).

The off-diagonal entries drive direct transitions from $|1\rangle$ to $|2\rangle$ with the two-photon Rabi frequency Ω_{R1} and from $|2\rangle$ to $|1\rangle$ with Ω_{R2}, where

$$\Omega_{Rn} = \frac{\Omega_{11}\Omega_{22}^*}{2\Delta_{nn}}, \quad n \in \{1, 2\}.$$ (2.20)

This results in a two-photon Rabi oscillation between the ground state and the metastable state, a transition that is optically forbidden in the dipole approximation. By performing two more unitary transformations the two-level Hamilton in Eq. (2.18) can be further simplified. The first one eliminates the diagonal elements by transforming $|\Psi\rangle_{IF_{od}} = U|\Psi\rangle_{IF}$, where

$$\hat{U} = \begin{pmatrix} e^{-i\Omega_1^{AC}} & 0 \\ 0 & e^{-i\Omega_2^{AC}} \end{pmatrix}$$ (2.21)

and the second one eliminates the remaining time dependencies by transforming $|\Psi\rangle_{2LVL} = U|\Psi\rangle_{IF_{od}}$ where

$$\hat{U} = \begin{pmatrix} e^{-i\frac{(\delta-\delta_{AC})}{2}t} & 0 \\ 0 & e^{+i\frac{(\delta-\delta_{AC})}{2}t} \end{pmatrix}.$$ (2.22)

with $\delta_{AC} = \Omega_1^{AC} - \Omega_2^{AC}$. This yields a time-independent two-level Hamiltonian

$$\hat{H}_{2LVL} = (-\hbar) \begin{pmatrix} (\delta - \delta_{AC})/2 & \frac{\Omega_{R2}}{2}e^{-i\phi_L} \\ \frac{\Omega_{R1}^*}{2}e^{i\phi_L} & -(\delta - \delta_{AC})/2 \end{pmatrix}.$$ (2.23)

Since all unitaries $\hat{U}$ that were used in the applied transformations are diagonal, no states were mixed, ergo the first vector component of $|\Psi\rangle_{2LVL}$ is still purely associated to the ground state $|1, p\rangle$.

2.2.2 Rabi Oscillations and Bloch Sphere

Rabi oscillations are used in many fields, amongst others in metrology, for the stabilization of atomic clocks via Ramsey-like interferometry [27]. Their description on the Bloch sphere has proven to be a powerful way to provide a comprehensive visualization of Rabi oscillations and coherent pulses. A more complete treatment of this topic can be found in textbooks such as [28]. Here it is just shown that a Raman transition can indeed be described as a Rabi oscillation.

Starting from equation (2.23) and switching to a description of the effective two-level system through a density matrix, the Liouville equation $i\hbar\frac{\partial\hat{\rho}}{\partial t} = [\hat{H}_{2\text{LVL}}, \rho]$, with $\hat{\rho} = \begin{pmatrix} \rho_{11} & \rho_{12} \\ \rho_{21} & \rho_{22} \end{pmatrix}$ yields

$$i\hbar\frac{\partial\rho_{11}}{\partial t} = -\frac{\hbar\Omega_{\text{R2}}}{2}e^{-i\phi_L}\rho_{21} + \frac{\hbar\Omega_{\text{R1}}^*}{2}e^{i\phi_L}\rho_{12}, \tag{2.24a}$$

$$i\hbar\frac{\partial\rho_{22}}{\partial t} = -\frac{\hbar\Omega_{\text{R1}}^*}{2}e^{i\phi_L}\rho_{12} + \frac{\hbar\Omega_{\text{R2}}}{2}e^{-i\phi_L}\rho_{21}, \tag{2.24b}$$

$$i\hbar\frac{\partial\rho_{12}}{\partial t} = \hbar\left(\delta - \delta^{AC}\right)\rho_{12} + \frac{\hbar\Omega_{\text{R2}}}{2}e^{-i\phi_L}(\rho_{11} - \rho_{22}), \tag{2.24c}$$

$$i\hbar\frac{\partial\rho_{21}}{\partial t} = -\hbar\left(\delta - \delta^{AC}\right)\rho_{21} - \frac{\hbar\Omega_{\text{R1}}^*}{2}e^{i\phi_L}(\rho_{11} - \rho_{22}). \tag{2.24d}$$

We can now define the Bloch vector as $\vec{R} = u\vec{e}_x + v\vec{e}_y + w\vec{e}_z$, where

$$u = \rho_{12} + \rho_{21}, \tag{2.25a}$$

$$v = -i(\rho_{21} - \rho_{12}), \tag{2.25b}$$

$$w = \rho_{22} - \rho_{11}, \tag{2.25c}$$

which is the projection of $\hat{\rho}$ onto the Bloch sphere in Cartesian coordinates. The time evolution of the components of the Bloch vector then reads

$$\frac{\partial u}{\partial t} = -\left(\delta - \delta^{AC}\right)v - \frac{1}{2}\left(\Omega_R e^{-i\phi_L} - \Omega_R^* e^{i\phi_L}\right)w, \tag{2.26a}$$

$$\frac{\partial v}{\partial t} = \left(\delta - \delta^{AC}\right)u + \frac{1}{2}\left(\Omega_R e^{-i\phi_L} + \Omega_R^* e^{i\phi_L}\right)w, \tag{2.26b}$$

$$\frac{\partial w}{\partial t} = i(\Omega_R^* e^{i\phi_L}\tilde{\rho}_{12} - \Omega_R e^{-i\phi_L}\tilde{\rho}_{21}). \tag{2.26c}$$

For $\delta \approx 0$ one can approximate $\Omega_{R1} = \Omega_{R2} =: \Omega_R$. Equation (2.26a)-(2.26c) can then be written as a vector equation

$$\frac{\partial \vec{R}}{\partial t} = \vec{R} \times \vec{\Omega}, \tag{2.27}$$

with $\vec{\Omega} = \Omega_R \cos(\phi_L)\vec{e}_x + \Omega_R \sin(\phi_L)\vec{e}_y + (\delta - \delta_{AC})\vec{e}_z$. Geometrically this is a rotation of $\vec{R}$ around $\vec{\Omega}$ with the frequency $\Omega = |\vec{\Omega}| = \sqrt{\Omega_R^2 + (\delta - \delta_{AC})^2}$.

This exactly corresponds to a single-photon Rabi oscillation between the states $|1\rangle$ and $|2\rangle$ with the Rabi frequency $\Omega_R = \frac{\Omega_{11}\Omega_{22}^*}{2\Delta}$, only that the single-photon resonance detuning Δ is replaced by the two-photon resonance detuning δ, corrected by the difference in AC-Stark shifts between state $|1\rangle$ and state $|2\rangle$.

In particular, this implies that for a system, in which initially only the ground state is populated, the population of the metastable state after interacting with both lasers for a given time t is given by

$$P_2(t) = \rho_{22}(t) = \frac{\Omega_R^2}{\Omega^2} \sin^2\left(\frac{\Omega}{2}t\right), \tag{2.28}$$

as can be geometrically derived on the Bloch sphere or using the dressed-state approach [29].

The amplitude of the Rabi oscillation is maximal in resonance ($\delta - \delta_{AC} = 0$), when $\frac{\Omega_R^2}{\Omega^2} = 1$. For optimal population transfer into the metastable state at a given interaction time t_{int},

$$\Omega_R \cdot t_{int} = (2n - 1)\pi \quad n \in \mathbb{N} \tag{2.29}$$

has to hold. This can be ensured by adjusting the interaction time as well as the laser power and detuning Δ since $\Omega_R \sim \frac{\sqrt{P_1 P_2}}{\Delta}$.

In the experiment, if the frequency of the second laser is fixed, and the frequency of the first laser is scanned, while probing the population of the metastable state. Thus, the expected line shape according to Eq. (2.28) is given by

$$P_2(\omega) = A\frac{\Omega_R^2}{\Omega_R^2 + (\omega_0 - \omega)^2} \sin^2\left(\frac{t_{int}}{2}\sqrt{\Omega_R^2 - (\omega_0 - \omega)^2}\right) + C \tag{2.30}$$

$$\approx A\frac{B}{B + (\omega_0 - \omega)^2} \sin^2\left(\frac{\Phi_0}{2} + \frac{\Phi_0}{4B}(\omega_0 - \omega)^2\right) + C, \tag{2.31}$$

where A is an amplitude, C a constant offset, $B = \Omega_R^2$ and $\Phi_0 = \Omega_R t_{\text{int}}$. δ_{AC} was absorbed into the resonance frequency ω_0 and in the second step the Taylor expansion $\sqrt{1+x} \approx 1 + x/2$ for $x \ll 1$ was used. This line shape corresponds to a Lorentzian that is modulated by a $\sin^2$-term.

2.2.3   Differences to the Full Hamiltonian

As mentioned above, the interaction of the first laser with transition 2 and the interaction of the second laser with transition 1 are missing in the Hamiltonian in Eq. (2.7c). The Hamiltonian including those interactions is given in [24], where it is also shown that (for $\Delta \gg \delta$) this indeed only leads to a second contribution in the AC-Stark shift (Eq. (2.19)). Spontaneous decay has been omitted. From the study of single-photon Rabi oscillations, one would expect spontaneous decay to lead to an exponential dampening of the Rabi oscillation with a limit of the form $\frac{1}{4}\Omega_R^2/(\frac{1}{2}\Omega_R^2+(\delta-\delta_{\text{AC}})^2+\frac{\Gamma^2}{4})$. Γ is some effective linewidth of the Raman transition that depends on the linewidths of transitions 1 and 2 as well as the metastable to ground state transition. The Liouville equation for the full three-level Hamiltonian including spontaneous decay is given in [14], however no further investigations have been done within this work.

More importantly, four aspects have not been considered yet:

Hyperfine structure: When studying transitions in nuclei with non-zero nuclear spin, the transitions will show hyperfine splitting. This results in possible Raman transitions between any triplet of $|1\rangle$, $|2\rangle$, and $|3\rangle$ hyperfine states that obey dipole selection rules for transitions 1 and 2. Additionally, when calculating dipole moments, Eq. (2.9) has to be adjusted to take into account the splitting of the total transition strength between the hyperfine states. A detailed treatment of this can be found in [26].

Velocity distribution: So far only ions of a fixed initial momentum and hence velocity have been considered. Under experiment-like conditions, ions have a velocity distribution, leading to a Doppler broadening of the Raman transition, see Sect. 2.3.1. A change in velocity also leads to a change in interaction time between laser and ions. At 20 kV the relative change in interaction time is roughly 25 ppm and therefore negligible for CLS.

Polarization: Depending on the laser polarization, the lasers will couple different magnetic substates with different strengths and the transitions between some magnetic substates are forbidden due to angular momentum conservation ($\Delta m = m' - m = -1, 0, +1$ for σ^-, π, σ^+ polarized light when absorbing a photon and $\Delta m = +1, 0, -1$ when emitting a photon, where m and m' are the magnetic quantum numbers of the initial and final state of the transition). This is reflected in different dipole moments for different m and m', a theoretical description of this can be found in [30].

Hence, the optimal laser power and detuning Δ required for a π-pulse is polarization dependent and there may be several maxima when maximizing the population of the metastable state. Changes in Rabi frequencies also lead to different AC-Stark shifts, which in theory leads to different peak positions for different polarizations.

Beyond the Λ-scheme: The deviation above can easily be expanded to a scenario, where the state $|2\rangle$ is a second excited state above the excited state $|3\rangle$. This requires a change in the sign of the definition of ω_{23} and a change in the sign of the momentum transfer of the second photon, since in this scenario both photons are absorbed. This will lead to a modified two-photon resonance condition $\Delta_{11} = -\Delta_{22}$, which fits the intuitive picture of a two-step excitation through a virtual intermediate state. In this case spontaneous decay from the $|2\rangle$ state should be considered since it is no longer metastable.

2.3 Linewidth Limitations of Single-Raman Collinear Laser Spectroscopy

The experimental approach in this work is to drive a Raman transition from the state $|1\rangle$ to the state $|2\rangle$ in a first interaction, of which the (rest-frame) laser frequencies are scanned. In a second, fixed, interaction, the population of the state $|2\rangle$ will be probed by resonantly driving the $|2\rangle \rightarrow |3\rangle$ dipole transition and measuring the photons emitted during the spontaneous $|3\rangle \rightarrow |1\rangle$ decay. The employed measuring schemes will be explained in detail in Sect. 3.5. As only the Raman transition, and not the dipole transition, is scanned, the measured linewidth is not influenced by the width of the dipole transition. However, different other factors lead to a broadening of the observed peak, which will be discussed next.

2.3.1 Doppler Broadening

The Doppler shifted laser frequencies in the ion rest-frame, and thus the laboratory-frame laser frequencies, under which the two-photon resonance condition is met, depend on the ion velocity. Hence, the finite width of the ion velocity distribution induces a broadening of the measured transition, known as Doppler broadening.

In collinear laser spectroscopy, Doppler broadening is caused by the velocity width Δv of the ion beam leaving the ion source. This velocity width is limited by the thermal movement of the ions. In the non-relativistic limit for a gas at rest following a Maxwell–Boltzmann velocity distribution, the resulting Doppler width $\delta\omega_D$ for a transition at the frequency ω_0 of ions at temperature T is given by

$$\delta\omega_D = \frac{\omega_0}{c}\sqrt{\frac{8k_B T \ln 2}{m}}, \tag{2.32}$$

with the Boltzmann constant k_B [21].

In collinear laser spectroscopy, Doppler broadening is reduced via Doppler compression. From the non-relativistic kinetic energy relation it follows that $\Delta v = \frac{\Delta E}{\sqrt{2mE}}$. Therefore, to reduce Doppler broadening, ions are accelerated to typically 10–60 keV [31], which compresses the Doppler width by 2–3 orders of magnitude, below the natural linewidth of allowed dipole transitions [1], typically in the 10–100 MHz range. Since the width of Raman transitions is in the kHz range, this does not suffice to suppress Doppler broadening when performing collinear Raman spectroscopy. A scheme to perform Doppler-free Raman spectroscopy will be discussed in Sect. 2.4.

To obtain the experimentally expected line shape, the line shape of the Raman transition in Eq. (2.30) has to be convoluted with the ion velocity distribution. Since Raman spectroscopy is phase sensitive and a change in velocity shifts Δ, and thus Ω_R, this may lead to a collapse of the Rabi oscillation [14]. The experimentally expected linewidth for a given energy width ΔE can be estimated as follows:

A shift ΔE in ion energy shifts the laser frequencies in the ion rest frame by $D(\omega_{L1}) \cdot \Delta E$ and $D(\omega_{L2}) \cdot \Delta E$ respectively, where

$$D(\omega_{L1/2}) = \frac{\partial \omega(\omega_{L1/2}, E, m, q)}{\partial E} \tag{2.33}$$

are the differential Doppler factors of Laser 1 and Laser 2, according to Eq. (2.1). To meet the two-photon resonance condition, the difference of both laser frequencies in the ion rest-frame must remain unchanged. The resonance condition for $\omega_{L1} - \omega_{L2}$ in the laboratory-frame hence shifts by $\Delta D \Delta E = (D(\omega_{L1}) - D(\omega_{L2}))\Delta E$, equivalent to a broadening by $\delta\omega_D = \Delta D \Delta E$.

Note that $D(\omega_{L1/2})$ decreases with E, reflecting the possibility of performing Doppler compression by accelerating the ion beam to higher energies, as mentioned above.

2.3.2 Time of Flight Broadening

If the interaction time t_{TOF} between lasers and ions is shorter than the lifetime $\tau = \frac{\hbar}{\Gamma}$ of the transition, the short interaction time leads to a broadening δ_{TOF} that is larger than the natural linewidth Γ of the transition [21].

The laser frequency spectrum A in the ion rest-frame is given by the Fourier transform of the electric field the ion experiences. For an ion interacting with a continuous-wave (CW) laser of frequency ω_L, neglecting the laser linewidth, for a time t_{TOF}, the frequency spectrum is given by

$$A(\omega) = \int_0^{t_{TOF}} E_0 \cos(\omega_0 t) e^{-i\omega t} dt. \tag{2.34}$$

For $(\omega - \omega_0) \ll \omega_0$ this results in an intensity distribution

$$I(\omega) = A \cdot A^* \sim \frac{\sin^2((\omega - \omega_0)t_{TOF}/2)}{(\omega - \omega_0)^2}. \tag{2.35}$$

The width of this distribution is

$$\delta\omega_{TOF} \approx 5.6/t_{TOF}. \tag{2.36}$$

In collinear laser spectroscopy, typical interaction times are in the μs-range, which induces a broadening in the 100 kHz-range. While negligible when performing collinear laser spectroscopy on dipole transitions, this has to be taken into account when investigating Raman transitions.

2.3.3 AC-Stark Broadening

Under experiment-like conditions, laser and ion beams have a spatial intensity distribution. While this has several implications that will be further discussed in Sect. 4.4, one shall already be mentioned here, as it impacts the observed line shape. The spatial intensity distribution of the laser beams implies that ions interacting with different parts of the lasers experience different electromagnetic field strengths. Since the Rabi frequencies Ω_{nj} and thus the AC-Stark shift depend on this field strength, different ions experience different AC-Stark shifts and thus have varying two-photon resonance conditions. This results in an additional broadening of the observed linewidth.

2.4   The Double Raman Scheme in a Four-Level System

Saturation spectroscopy is an established method to perform Doppler-free spectroscopy, which has already been tested at COALA [32]. In a first step, ions of one velocity class are pumped into an excited state with a laser that is kept a a constant laser frequency. This results in a so-called Lamb dip in the population of the ground state. The width of the Lamb dip is not Doppler broadened and depends chiefly on the natural linewidth, the laser linewidth and power broadening. If now the same transition is probed via a second laser, additional ions are excited, increasing the population of the excited state. The velocity of the ions excited in the second interaction depends on the frequency of the second laser. If this velocity matches the velocity of the ions excited in the first interaction, fewer ions are excited in the second interaction, since the transition has been saturated in the first interaction. This results in a Lamb dip in the resonance signal, allowing for Doppler-free spectroscopy.

If the transition is not closed, *e.g.*. if the excited state partly decays into a metastable state, alternatively to using a second laser, the second interaction can also be performed via Doppler tuning in a second interaction region.

In this work, the option of expanding this concept to Raman transitions is explored. However, since a Raman transition is a two-photon Rabi oscillation, see Eq. 2.28, a few aspects have to be taken into account: If the velocity of the ions in resonance in the second interaction matches the velocity of the ions selected in the first interaction, not only are less additional ions transferred into the excited state, but also are ions of that velocity transferred back into the ground state. The latter enhances the dip in the resonance signal. Secondly, unlike for a dipole transition, the population transfer of a Raman transition does not monotonically increase with the laser intensity. This becomes especially relevant if spatial intensity distributions are considered, see Sect. 4.4.

If the system of interest has more than one metastable state, this opens up the possibility to perform a separate second sequential Raman transition. Figure 2.2 shows such a four-level system. The same notation as in Sect. 2.2 Fig. 2.1 is used.

However, the intermediate state is now $|4\rangle$, and $|3\rangle$ is the second metastable state. A third laser (frequency ω_{L3}, wave vector k_{L3}) is introduced that couples $|3\rangle$ and $|4\rangle$ via a dipole transition with the Rabi frequency Ω_{33}.

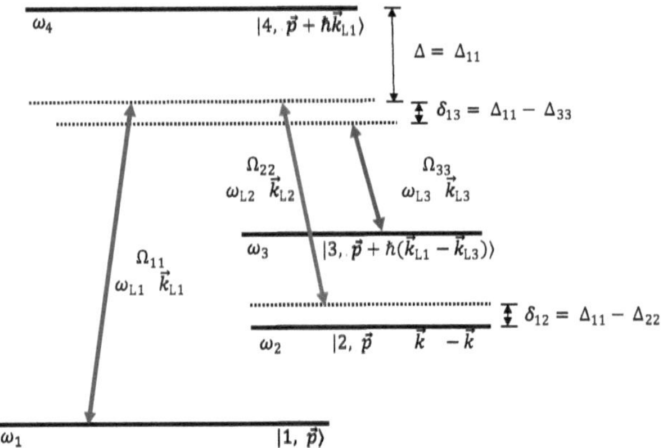

Fig. 2.2 Four-level system with a ground state $|1\rangle$ and two metastable states $|2\rangle$ and $|3\rangle$. The arrows indicate the coupling of the laser $j \in \{1, 2, 3\}$ (frequency ω_{Lj}, wavevector $\vec{k}_{Lj}$) with the transition between level $|n\rangle$, $n \in \{1, 2, 3\}$ and the intermediate short-lived state $|4\rangle$ with the Rabi frequency Ω_{nj} and the single-photon detuning Δ_{nj}. Off-resonant couplings (see Fig. 2.1) are not shown

Two new possibilities arise: Instead of driving the same Raman transition from the ground state $|1\rangle$ into the metastable state $|2\rangle$ in the second interaction as in the first interaction, a Rabi transition from the ground state $|1\rangle$ into the second metastable state $|3\rangle$ can be driven. A measurement of this transition will still exhibit a Lamb dip.

Secondly, if both metastable states $|2\rangle$ and $|3\rangle$ are connected to the same intermediate state $|4\rangle$ through a dipole transition, as is the case in the system considered here, a Raman transition from $|2\rangle$ to $|3\rangle$ can be driven, as shown in Fig. 2.3. The energy width of the population of $|2\rangle$ is the unbroadened width of the first Raman transition. Thus, this allows a direct Doppler-free measurement of the transition frequency ω_{23} between the two metastable states.

For both scenarios, the two Raman transitions have to be performed sequentially. Otherwise, the ground state population is transferred to both metastable states, and at the same time, the population of both metastable states is transferred to the other metastable state, as well as back into the ground state, resulting in a loss of signal.

Energy conservation dictates, that $\omega_{12} + \omega_{23} = \omega_{13}$. Hence, if all three Raman transitions ($|1\rangle \rightarrow |2\rangle$, $|1\rangle \rightarrow |3\rangle$, and $|2\rangle \rightarrow |3\rangle$) are measured, a ring closure can be used to check and constrain experimental results.

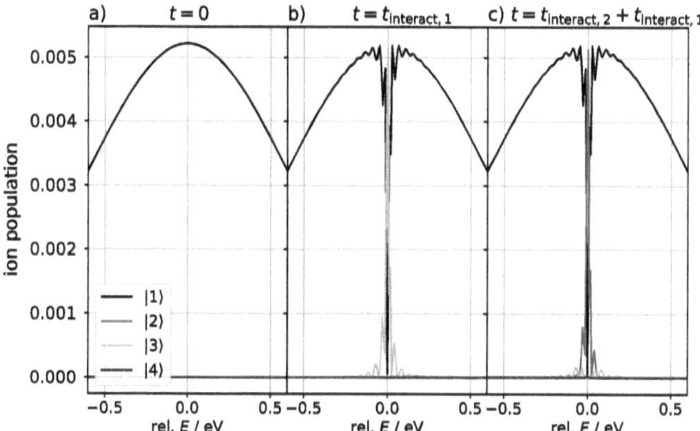

Fig. 2.3 Energy distribution in the different states of the four-level system, at different stages of the double Raman scheme. The ion energy was assumed to be normally distributed with a standard deviation of 500 meV.
a) Initially ($t = 0$) all ions are in the ground state $|1\rangle$.
b) In a first step ions are transferred at into the state $|3\rangle$ through a Raman transition with the interaction time $t_{\text{interact},1}$. This creates a Lamb dip in the ground state and populates the state $|3\rangle$ with an energy width equal to the width of the first Raman transition.
c) A second Raman transition from the state $|3\rangle$ to the state $|2\rangle$ (interaction time $t_{\text{interact},1}$) can then be probed Doppler-free

Analogous to Eq. 2.11, such a four-level system interacting with three lasers (see Fig. 2.2), can be described using the following Hamiltonian in the interaction frame:

$$\hat{H}_{\text{IF}} = \begin{pmatrix} 0 & 0 & 0 & \frac{\hbar\Omega_{11}}{2} e^{-i(\Delta t + \phi_1)} \\ 0 & 0 & 0 & \frac{\hbar\Omega_{22}}{2} e^{-i((\Delta + \delta_{12})t + \phi_2)} \\ 0 & 0 & 0 & \frac{\hbar\Omega_{33}}{2} e^{-i((\Delta + \delta_{13})t + \phi_3)} \\ \frac{\hbar\Omega_{11}^*}{2} e^{i(\Delta t + \phi_1)} & \frac{\hbar\Omega_{22}^*}{2} e^{i((\Delta + \delta_{12})t + \phi_2)} & \frac{\hbar\Omega_{33}^*}{2} e^{i((\Delta + \delta_{13})t + \phi_2)} & 0 \end{pmatrix} \tag{2.37}$$

Where $\Delta = \Delta_{11}$, $\delta_{12} = \Delta_{11} - \Delta_{22}$, and $\delta_{13} = \Delta_{11} - \Delta_{33}$. Δ_{nj} is the detuning of the laser j (frequency ω_{Lj}, wave vector k_{Lj}) from the dipole transition between the state $|n\rangle$ and $|4\rangle$ with the rest-frame transition frequency ω_{n4}, see Eq. 2.12 and 2.12 (note the index changed for the intermediate state).

To simplify the calculation of the time-dependent state population, time dependencies can be eliminated from Eq. 2.37, by applying a unitary transformation with

$$\hat{U}_{t\,\text{ind}} = \begin{pmatrix} 1 & 0 & 0 & 0 \\ 0 & e^{i\delta_{12}t} & 0 & 0 \\ 0 & 0 & e^{i\delta_{13}t} & 0 \\ 0 & 0 & 0 & e^{-i\Delta t} \end{pmatrix}. \tag{2.38}$$

resulting in

$$\hat{H}_{t\,\text{ind}} = \begin{pmatrix} 0 & 0 & 0 & \frac{\hbar\Omega_{11}}{2}e^{-i\phi_{L1}} \\ 0 & -\hbar\delta_{12} & 0 & \frac{\hbar\Omega_{22}}{2}e^{-i\phi_{L2}} \\ 0 & 0 & -\hbar\delta_{13} & \frac{\hbar\Omega_{33}}{2}e^{-i\phi_{L3}} \\ \frac{\hbar\Omega_{11}^*}{2}e^{i\phi_{L1}} & \frac{\hbar\Omega_{22}^*}{2}e^{i\phi_{L2}} & \frac{\hbar\Omega_{33}^*}{2}e^{i\phi_{L3}} & \hbar\Delta \end{pmatrix}. \tag{2.39}$$

This allows calculating the time evolution of an initial state $|\Phi\rangle_0 = |\Phi\rangle(t = 0)$ using the ansatz

$$|\Phi\rangle(t) = e^{-i\hbar\hat{H}_{t\,\text{ind}}t}. \tag{2.40}$$

2.5 Strontium Level Scheme

All measurements within this work were done in the $^2S_{1/2}$-$^2P_{3/2}$-$^2D_{3/2}$/$^2D_{5/2}$ and Λ-schemes of singly charged strontium (Sr, $Z = 38$), which is shown in Fig. 2.4. The states are denoted $^{2S+1}L_J$. S is the spin quantum number, $L = S, P, D$ for $L = 0, 1, 2$ is the orbital quantum number in spectroscopic notation and J is the total angular momentum.

In Sr$^+$ all subshells up to the $4p$ subshell are filled and, like for alkali metals, the electron configuration of low-lying states is defined by a single valence electron, hence, the spin quantum number is always $1/2$. Sr$^+$ therefore is an excellent candidate to benchmark atomic structure calculation, since one-valence-electron systems can be calculated with low theoretical uncertainty.

Strontium was chosen since its electron configuration is very similar to the one investigated in calcium, for which detailed calculations on a Raman velocity filter were conducted [14]. Additionally, Sr$^+$ has recently been investigated with high precision at COALA [33].

For the ground state to virtual state transition, the 408-nm $S_{1/2} \rightarrow P_{3/2}$ D2 transition was chosen and for the metastable state to virtual state, the 1004-nm

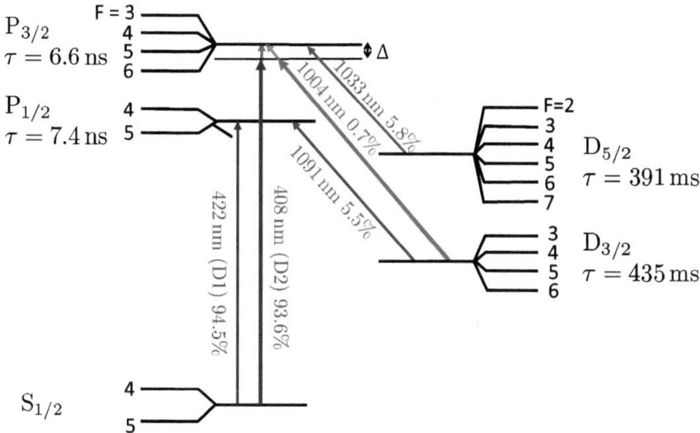

Fig. 2.4 The level scheme of Sr^{+} with all transitions of the ^{2}S-^{2}P-^{2}D Λ-scheme, including fine structure splitting. Shown are the wavelength of each transition (see Table 2.1) as well as the branching ratios and the lifetime for each excited state. In addition, the hyperfine splitting is shown for ^{87}Sr ($I = 9/2$). The transitions selected for the Raman transition are highlighted and the virtual intermediate state is indicated as a dashed line

D$_{3/2}$ $\rightarrow$ P$_{3/2}$ transition and the 1033-nm D$_{5/2}$ $\rightarrow$ P$_{3/2}$ transition were chosen, since wavelengths higher than 1050 nm were not accessible with the available laser setup. As it is the most abundant isotope, ^{88}Sr was selected.

The parameters required to simulate collinear Raman spectroscopy in the strontium S$_{J}$-P$_{J}$-D$_{J}$ Λ-scheme are summarized in Table 2.1 and 2.2. Table 2.1 shows

Table 2.1 Overview of abundances, masses, nuclear spins I, and absolute transition frequencies of the stable Sr isotopes. The transition frequencies $\nu_{L_J \rightarrow L'_{J'}}$ were taken from [33], except for $\nu_{D_{3/2} \rightarrow P_{3/2}}$, here the given values are preliminary values from recent measurements at COALA, their uncertainties might be underestimated

Isotope	^{88}Sr	^{87}Sr	^{86}Sr	^{84}Sr
Abundance (%) [34]	82.58(35)	7.00(20)	9.86(20)	0.56(2)
Mass (μu)[35]	87905612.254(6)	86908877.495(6)	85909260.725(6)	83913419.1(13)
I [36]	0	9/2	0	0
ν_{D1} (MHz)	710 962 838.07(17)	710 962 781.14(17)	710 962 666.56(17)	710 962 462.85(18)
ν_{D2} (MHz)	734 989 824.27(17)	734 989 768.43(17)	734 989 653.51(17)	734 989 449.48(17)
$\nu_{D_{3/2} \rightarrow P_{1/2}}$ (MHz)	274 589 146(30)	274 589 331(30)	274 589 540(30)	274 589 970(30)
$\nu_{D_{5/2} \rightarrow P_{3/2}}$ (MHz)	290 210 798(30)	290 210 990(30)	290 211 199(30)	290 211 621(30)
$\nu_{D_{3/2} \rightarrow P_{3/2}}$ (MHz)	298 616 134(30)	298 616 323(30)	298 616 535(30)	298 616 959(30)

Table 2.2 List of Einstein coefficients A in Sr^+, data taken from [37]

Transition	D1	D2	$P_{1/2} \to D_{3/2}$	$P_{3/2} \to D_{5/2}$	$P_{3/2} \to D_{3/2}$
$A\ (10^6\,\frac{1}{s})$	127.9	141	7.46	8.7	1.0

the masses of all stable Sr isotopes, their abundances, nuclear spins and rest frame transition frequencies. Table 2.2 shows the Einstein coefficients of the transitions shown in the level scheme in Fig. 2.4.

Experimental Setup   3

The COALA (<u>Co</u>llinear <u>A</u>pparatus for <u>L</u>aser Spectroscopy and <u>A</u>pplied Science) beamline at the Institute for Nuclear Physics at TU Darmstadt was designed to perform collinear laser spectroscopy on stable isotopes as well as to investigate novel experimental techniques [31], allowing for a level of precision previously unmatched in the field of collinear laser spectroscopy [38, 39].

The following chapter will provide an overview of the current state of the COALA setup, as well as the laser systems and the ion source used for the measurements performed within this work.

3.1   Laser Setup

As described in Sect. 2.5, the chosen Raman schemes require a 1004-nm and a 408-nm laser for the $S_{1/2} \rightarrow D_{3/2}$ transition and a 1033-nm and a 408-nm laser for the $S_{1/2} \rightarrow D_{5/2}$ transition. In this experiment, two separate laser systems placed in a dedicated laboratory next to the COALA beamline were used. They are schematically depicted in Fig. 3.1.

Each system consists of a Sirah Matisse 2 TS Ti:Sa laser, tunable from 660 nm to 1000 nm, that is pumped by a 25 W Millennia eV diode-pumped, and frequency-doubled Nd : YVO$_4$ laser.

The frequency selection within the Matisse laser is realized by a birefringent filter, two etalons, and a piezo-mounted adjustable mirror. Both Ti:Sa lasers are stabilized to a reference cavity, which is long-term stabilized to a Menlo Systems FC1500-250-WG frequency comb. A HighFinesse WS8-2 wavemeter, calibrated

© The Author(s), under exclusive license to Springer Fachmedien Wiesbaden GmbH,   23
part of Springer Nature 2026
J. Spahn, *Pioneering Raman Transition Techniques in Collinear Laser Spectroscopy for High-Precision Experiments*, BestMasters,
https://doi.org/10.1007/978-3-658-50605-6_3

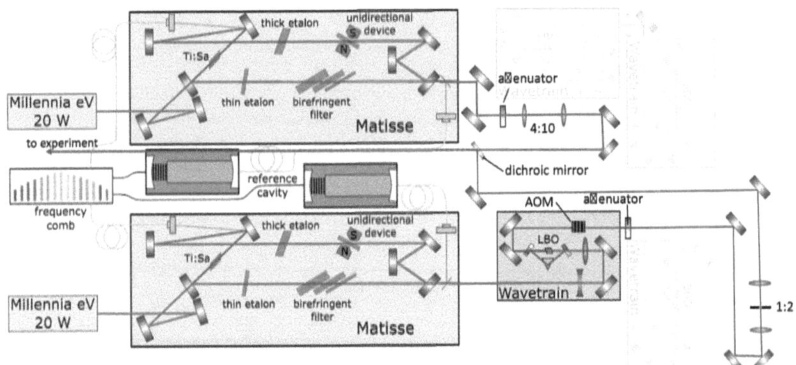

Fig. 3.1 Schematics of the laser systems at COALA; the upper system is used to produce the 1004-nm laser beam and the lower half system is used to produce the 408-nm laser beam. Additional frequency doubling stages available at COALA to produce down to 200-nm light, not used for this experiment, are grayed out. The figure was adapted from [31]

against a stable He:Ne-Laser, is used in combination with the frequency comb, to monitor the laser frequencies.

One of the Ti:Sa lasers is used to directly produce 1004-nm or 1033-nm light. Latter is achieved by installing a Sirah MOS3 extender mirror into the Matisse, allowing tuning of the Matisse up to 1050 nm. A typical laser power of 1 W is achieved at 16 W pumping power. The laser is then guided through an attenuator, to adjust the laser power used for the experiment. A 4:10 telescope is used to collimate the laser beam.

Since 408-nm light can not directly be produced from a Ti:Sa laser, 816-nm light is produced instead and then fed into a frequency doubler (Sirah wavetrain with a lithium triborate (LBO) crystal) to obtain a laser power of up to 300 mW at 408 nm (Millennia eV at 12 W).

Driving a Raman transition requires roughly one order of magnitude more laser power than driving an optical dipole transition. Since the PMTs used to detect fluorescence light are very sensitive to UV light, an Acousto-Optical Modulator (AOM) of the type MQ200-A1,5-244.266-B from AA-Opto-Electronic [40] is installed in the frequency doubling stage, to quickly turn the laser on and off, allowing for background-free spectroscopy, as described in Sect. 3.5.1. Since the beam diameter diverges after the Wavetrain, installing the AOM inside the Wavetrain minimizes the rise and fall time of the AOM.

Again an attenuator is used to adjust the laser power and the laser beam is collimated using a 1:2 telescope. To optimize the beam profile and reduce the laser-induced PMT background, a 40μm pinhole is placed in the focal plane of the telescope.

The 408-nm beam is superimposed with the 1004-nm or 1033-nm beam using a long-pass dichroic mirror. The collimated and superimposed laser beams are then guided through the COALA beamline in anticollinear direction using several mirrors. While this setup is more complicated than using an optical fiber, it allows to surpass limitations in laser power imposed by the use of optical fibers.

To perform quasi-simultaneous collinear/anticollinear measurements on the $S_{1/2} \rightarrow D_{5/2}$ transition and for measurements in the double Raman scheme, a Toptica DL Pro tuneable diode laser at 1033 nm, provided by the group of Ruben de Groote from KU Leuven, is used as the third laser. It is stabilized to a High Finesse WSU 30 wavemeter via a PID-loop and monitored with the WS8-2 wavemeter. The WSU 30 is calibrated to the same He:Ne-laser as the WS8-2. Stabilizing to the WS8-2 is not possible, since the expansion card providing the analog PID signal is only available for the WSU 30 wavemeter. The Diode laser is coupled into an optical fiber and used in collinear direction.

As the two Matisse lasers are stabilized to the frequency comb and the frequency comb can only measure two beats at once, the diode laser can not be monitored using the frequency comb during measurements. Instead, calibration measurements are performed to determine the offset of the WS8-2 wavemeter, as described in Sect. 6.4.1).

3.2 Frequency Comb

The WS8-2 wavemeter has a specified absolute 3σ-uncertainty of 30 MHz, which limits the precision of absolute frequency measurements. Investigations have shown, that this uncertainty is dominated by a frequency dependent offset, that changes by less than 50 kHz/h [41].

Frequency combs do not rely on fits to an interference pattern to measure frequencies, but on electronically counted beat signals. This allows to improve the accuracy of frequency measurements and correct for the wavemeter offset.

A 1550-nm femtosecond laser creates a comb-like frequency spectrum. The modes of this spectrum are spaced by the repetition rate $f_{\text{rep}} = 250$ MHz of the femtosecond laser and have an offset that is given by the carrier-envelope-offset (CEO) frequency f_{CEO}. This results in a frequency f_n of the nth comb mode of

$$f_n = f_{\text{CEO}} + n \cdot f_{\text{rep}}$$

f_{CEO} is measured by frequency doubling the amplified comb-light and overlapping the upper end of a comb octave with the lower end of the same, frequency-doubled octave, see Fig. 3.2. Since $2(f_n) - f_{2n} = f_{\text{CEO}}$ this results in a beat signal at the CEO frequency (CEO beat), which is stabilized to 20 MHz.

The comb light is first frequency doubled and then guided through a nonlinear PCF (Photonic Crystal Fiber), broadening and shifting the comb spectrum to a range between 680 nm and 1050 nm. The comb light is then separated into its frequency components using an optical grating and the frequency component, matching the laser frequency f_{CW} that is to be measured, is coupled into photo diode. The laser to be measured is coupled into the same diode. The beat frequency f_{beat} between the comb and the laser is electronically counted and stabilized to $f_{\text{beat}} = 60$ MHz. The frequency measurements of f_{rep}, f_{CEO} and f_{beat} are referenced to a 10-MHz GPS-disciplined quartz oscillator.

Since the sign of both beat signals is not known, an unambiguous frequency assignment is not directly possible. Since the four possible solutions are separated by more than 30 MHz, the wavemeter is used to identify the correct signs of the beats, enabling frequency measurements at a sub 50 kHz accuracy, primarily limited by the laser linewidth [41]. The frequency f_{CW}, to which the laser is stabilized, can be adjusted by varying f_{rep}.

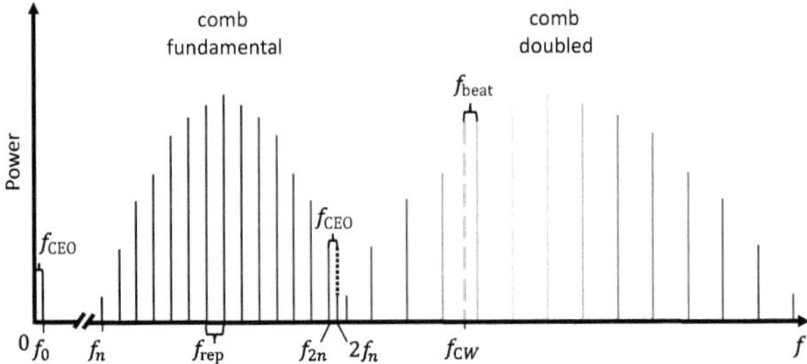

Fig. 3.2 Working principle of the frequency comb. The CEO beat frequency f_{CEO} is stabilized to 20 MHz and the beat f_{beat} between the comb and the continous-wave laser of interest is stabilized to 60 MHz. The repetition rate of the femtosecond laser f_{rep} is locked via GPS. This allows measuring the laser frequency f_{CW} with an uncertainty below 50 kHz

3.3 Surface Ionization Source

At the COALA beamline, different sources are available for ion production. Recently used sources include an Electron Beam Ion Source (EBIS) [39, 42], a laser ablation source [43] and, a penning ionization gauge source [44].

To produce Sr^+ ions a surface ionization source, built by N. Frömmgen [45] and based on the design by S. Raeder [46], was used. A graphite crucible was filled with strontium and was heated with an electric current of up to 85 A, evaporating the strontium. The source was updated with a tantalum heat shield to improve efficiency and temperature. Figure 3.3 compares the power-temperature curve measured by N. Frömmgen [45] to measurements done at KU Leuven [47] on the same source design with a heat shield, using an infrared camera. The efficiency is improved by 40–50% and higher maximum temperatures might be achievable when supplying more than 150 W. In this work, typical currents of 43 A, corresponding to 150 W, were used, heating the source to 2200 °C via Ohmic heating. Since graphite has a work function of about 5 eV [45] and the first ionization energy of Sr is 5,7 eV [37], an electron from the Sr atom can tunnel into the graphite when hitting the graphite crucible. This method of ionization is known as surface ionization or thermal ionization.

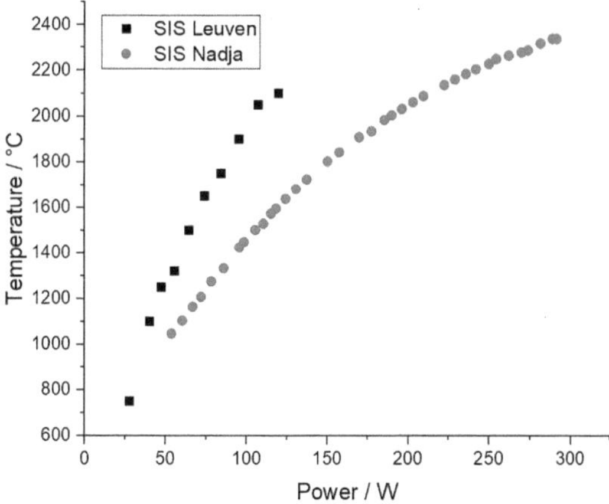

Fig. 3.3 Temperature of the SIS depending on the applied heating power. The red dots are measurements in the original design, done by N. Frömmgen [45], and the black squares are measurements with an additional tantalum heat shield around the graphite crucible [47]. The latter was used in the present setup

3.4   COALA Beamline

In this section, an overview of the current setup of the 7-m long COALA beamline, depicted in Fig. 3.4, is provided, following the trajectory of the ion beam.

The surface ionization source is placed on a high-voltage potential U_{acc} of 20 kV, which is applied with a Heinzinger PNChp 20000-10 high-voltage power supply and monitored using a high-voltage divider and a Keysight 3458A multimeter. Small drifts of the voltage are compensated using the multimeter readout and a custom DAC.

The ions are then accelerated to ground potential, focused using an einzel lens, and guided into the switchyard by a pair of vertically and a pair of horizontally mounted steerer electrodes (steerer 1). Inside the switchyard [42], two of the six steerer electrodes are used to bend the ion beam by 10° into the main beamline.

There, the ion beam is guided through a quadrupole doublet to optimize the ion beam shape and a second set of steerer electrodes (steerer 2) is used, in combination with the other ion optics, to align the ion beam with the laser beams.

Three beam diagnostic stations along the beamline are used to monitor the ion beam shape as well as the overlap between the laser and ion beams. They consist of an adjustable iris aperture and a Faraday cup that can be lowered into the beam to measure the ion current. Additionally, the first beam diagnostic station is equipped with a mirror facing the ion source, that can also be lowered into the beamline, to do a coarse adjustment of the ion source position. The second and third beam diagnostics are equipped with a transparent phosphorus screen connected to a multi-channel plate instead of a mirror, to detect both the ion beam and the laser beams in collinear and anticollinear direction.

After the second beam diagnostic station, a 1.2-m long pumping tube can be floated to a separated potential $U_{pump,1}$ for optical pumping. In this experiment, this tube was used as an interaction region for the Raman transition, since it has a fixed potential. This would not be the case when using the entirety of the beamline, due to the voltage applied to the ion optics and contact potential between different components of the beamline. The einzel lens placed after the optical pumping tube was used as a second, 0.3-m long, interaction region for the second Raman transition of the double Raman scheme by floating all three electrodes to a second interaction potential $U_{pump,2}$.

To apply $U_{pump,1}$ and $U_{pump,2}$, each interaction region is connected to a Kepco BOP500M voltage amplifier with an amplification factor of 50. The voltage amplifiers are supplied with a DAC voltage between -10 V and 10 V.

The Fluorescence Detection Region (FDR) at the end of the beamline consists of two sections that focus fluorescence light into Photo Multiplier Tubes (PMTs), to perform fluorescence spectroscopy. In the first section, two elliptical mirrors are

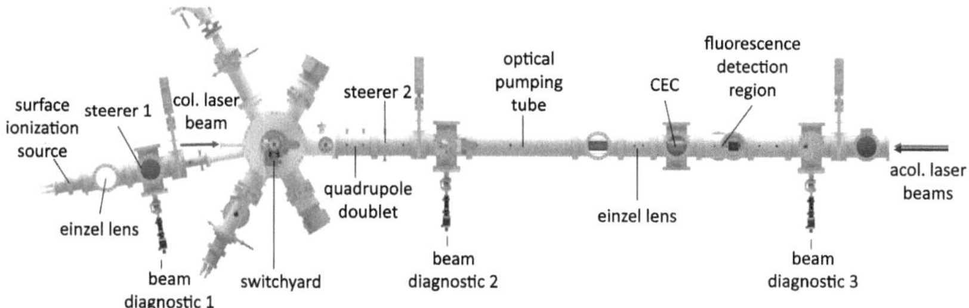

Fig. 3.4 Setup of the COALA beamline at TU Darmstadt as described in Sect. 3.4. The Sr^+ beam is produced in the surface ionization source and superimposed with the two anticollinear and the collinear laser beams in the switchyard. The overlap of the beams can be monitored in the diagnostic stations and is optimized with further ion optical elements. The Raman excitation takes place in the 1.2-m long optical pumping tube and the ions are finally probed in the fluorescence detection region, where fluorescence photons can be collected with photomultiplier tubes. Optionally, a second Raman transition can be driven in the einzel lens between the optical pumping tube and the charge exchange cell (CEC)

placed along the beam axis to collect and guide the fluorescence light into two PMTs while in the second chamber, a lens-based system is used to focus the fluorescence light into a single PMT [39].

The PMT signals are extended using a custom pulse extender to be accepted by the time-resolved data acquisition system TILDA [48], built on the National Instruments XPI platform using two field-programmable gate arrays (FPGA), which also generate the signal used to remotely scan the laser frequency or the voltage applied to the interaction region and trigger the AOM for background-free measurements.

3.5   Measurement Scheme

3.5.1   Background-Free Single Raman

The typical measurement scheme for single-Raman spectroscopy is depicted in Fig. 3.5. Driving for example the $S_{1/2} \rightarrow D_{3/2}$ Raman transition leads to an increase in the $D_{3/2}$ state population. Probing the $D_{3/2}$ state population thus allows the detection of the population transfer induced by the Raman transition. This was achieved by resonantly probing the $D_{3/2} \rightarrow P_{3/2}$ transition. Ions resonantly excited into the $P_{3/2}$ state spontaneously decay into the $S_{1/2}$ ground state, emitting 408-nm photons. Those can then be detected by the PMTs. As the number of emitted 408-nm photons is proportional to the population of the $D_{3/2}$ state, this enables detecting the Raman transition.

Driving the $S_{1/2} \to D_{3/2}$ Raman transition implies that the 1004-nm laser is not on resonance and can thus not directly be used to probe the $D_{3/2}$ state. To compensate for this detuning of the 1004-nm laser, a fixed voltage is applied to the FDR, using an ISEG high-voltage power supply. This shifts the 1004-nm rest-frame frequency via Doppler tuning into resonance. In particular, no third laser is needed to measure the Raman transition.

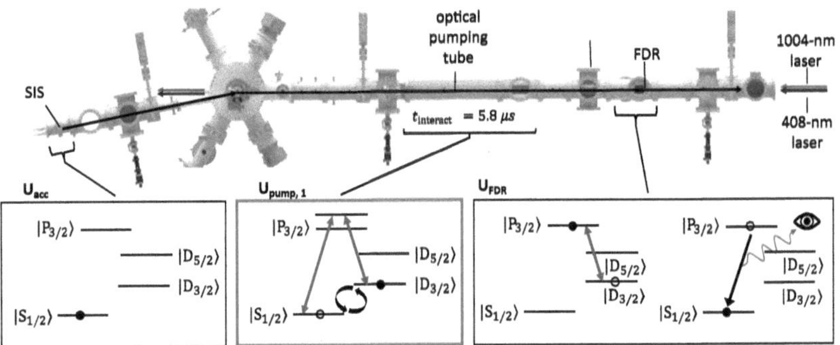

Fig. 3.5 Measurement scheme employed to perform background-free collinear Raman spectroscopy, illustrated for the $S_{1/2} \to D_{3/2}$ Raman transition. The electronic state populations for the different steps are indicated by black circles in the level schemes. Black bended arrows indicate transitions between the different states and red/blue straight arrows indicate the 1004-nm and 408-nm laser, respectively.
Ions leaving the SIS, which is placed at an acceleration potential U_{acc}, are (up to the negligible thermal population of the meta-stable states) in the $S_{1/2}$ ground state. The optical pumping tube is floated to a potential $U_{\mathrm{pump},1}$. Spectroscopy on the Raman transition is performed via Doppler tuning by scanning $U_{\mathrm{pump},1}$. In a second step, the $D_{3/2}$ state is probed by resonantly addressing the $D_{3/2} \to P_{3/2}$ transition and detecting the photons emitted from the $P_{3/2} \to S_{1/2}$ decay. During this step, the 408-nm laser is turned off to eliminate the laser-induced background

The background induced by the 408-nm laser is a limiting factor when pushing to higher laser powers, due to saturation of the PMTs, or when the population transfer into the metastable state is not optimal. To remove the background induced by the 408-nm laser, background-free spectroscopy was implemented. For this purpose, the following two steps are repeated multiple times per scan step:

1. Turning the 408-nm laser on and waiting long enough for the ions to fly through the interaction region at a potential $U_{\mathrm{pump},1}$, where they get transferred into the metastable state via the $S_{1/2} \to D_{3/2}$ Raman transition, driven by the 408-nm and 1004-nm laser. The 408-nm laser is kept on until the excited ions reach the

FDR to optimize the efficiency of the duty cycle, given by the ratio between laser-on and laser-off time.

2. Turning the 408-nm laser off and probing the population of the metastable $D_{3/2}$ state by resonantly driving the $D_{3/2} \rightarrow P_{3/2}$ with the still active 1004-nm laser, and detecting the 408-nm photons from the spontaneous $P_{3/2}$ state to ground state decay inside the FDR. For this, the detuning of the 1004-nm laser, at which the Raman transition is driven, is compensated by applying a voltage U_{FDR} to the FDR.

As described in Sect. 3.1 and Sect. 3.4, an AOM was used to switch the laser on and off, and the pattern generator of the TILDA data acquisition system was used to generate the AOM trigger signal. From the distance of $1,8\,\text{m}$ from the beginning of the interaction region to the FDR, a laser-on-time of $8.7\mu\text{s}$ was estimated. As the time of flight through the FDR is $0.9\mu\text{s}$, an off-time of $10.6\mu\text{s}$ was chosen. While decreasing the efficiency of the scheme, with this high off-time, ions that only partially interacted with the 408-nm laser in the optical pumping tube during the laser-on time, are also probed. This enables the investigation of two-photon Rabi oscillations, which will be further elaborated in Sect. 5.4.

The Raman resonance is measured via Doppler tuning by measuring the PMT counts, while scanning $U_{\text{pump},1}$. The voltage applied to the FDR is kept constant, to continuously probe the $D_{3/2}$ state.

Analogously, the $S_{1/2} \rightarrow D_{5/2}$ can be measured by probing the $D_{5/2}$ state using the 1033-nm laser.

3.5.2 Background-Free Double Raman

To realize the Doppler-free double Raman scheme outlined in Sect. 2.4, the measurement scheme for single Raman transitions is expanded as follows: Instead of scanning the optical pumping tube, the first Raman transition is driven by fixing $U_{\text{pump},1}$ on the Raman resonance. This results in a Lamb dip in the ground state population and a metastable population with an energy width corresponding to the width of the first Raman transition. A second pumping region is introduced and floated to the voltage $U_{\text{pump},2}$. If an $S_{1/2} \rightarrow D_{5/2}$ Raman transition is driven in the first interaction region, as shown in Fig. 2.4, three options arise:

1. Scanning $U_{\text{pump},2}$ over the $S_{1/2} \rightarrow D_{5/2}$ transition at $U_{\text{pump},1}$ to measure the $S_{1/2} \rightarrow D_{5/2}$ Raman resonance with Lamb dip. Analogous to Sect. 3.5.1, the $D_{5/2}$ state is probed through the $D_{5/2} \rightarrow P_{3/2}$ transition by floating the FDR to a

voltage U_{FDR} to compensate for the detuning of the 1033-nm laser. The photons from the[2.] $P_{3/2} \rightarrow S_{1/2}$ decay are then counted by the PMTs.

2. Scanning $U_{\text{pump},2}$ over the $S_{1/2} \rightarrow D_{3/2}$ transition, driven by the 408-nm laser and a third laser at 1004 nm. Driving this transition at a voltage $U_{\text{pump},2} \neq U_{\text{pump},1}$ ensures that the first and the second interaction are well separated, see Sect. 2.4, by shifting the 408-nm frequency in the ion rest-frame between the $S_{1/2} \rightarrow D_{5/2}$ and the $S_{1/2} \rightarrow D_{3/2}$ transition. This also implies that the two transitions are driven at a different detuning. As the first interaction imprints a Lamb dip in the $S_{1/2}$ ground state population, this measurement will also exhibit a dip. Like in Sect. 3.5.1, a voltage U_{FDR} is applied to the FDR to probe the $D_{3/2}$ state via the $D_{3/2} \rightarrow P_{3/2}$ transition.

3. Doppler-free measurement of the $D_{5/2} \rightarrow D_{3/2}$ transition, driven by the 1033-nm laser and the 1004-nm laser. The two interactions are separated by driving the second interaction at $U_{\text{pump},2} \neq U_{\text{pump},1}$, changing the detuning of the 1033-nm laser between both interactions. Again, the $D_{3/2}$ population is probed through the $D_{3/2} \rightarrow P_{3/2}$ transition inside the FDR.

In Fig. 3.6 the third case, a Raman transition into the $D_{5/2}$ state and a consecutive second transition from the $D_{5/2}$ to the $D_{3/2}$ state, is illustrated. Separating the first and second transition by shifting the rest-frame laser frequencies via Doppler tuning in particular bypasses the need of a forth laser.

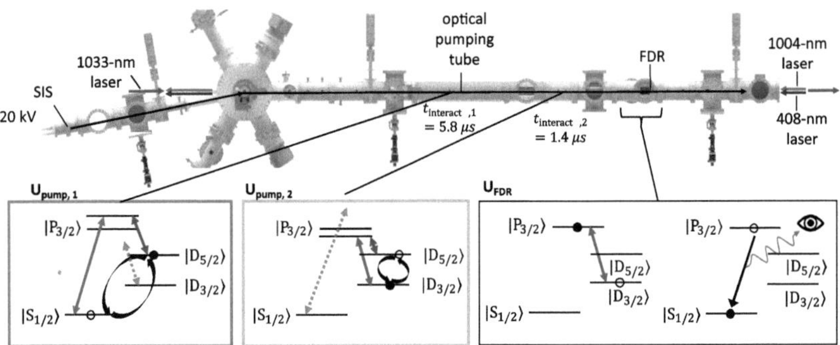

Fig. 3.6 Measurement scheme employed to perform Doppler-free collinear Raman spectroscopy, shown for a $S_{1/2} \rightarrow D_{5/2}$ Raman transition in the optical pumping tube (voltage $U_{\text{pump},1}$) and a sequential $D_{5/2} \rightarrow D_{3/2}$ Raman transition in the einzel lens, which is used as second pumping region by floating all three electrodes to the voltage $U_{\text{pump},2}$. The black circles mark the population at different steps of the measurement scheme. Level transitions are indicated by bended black arrows, while solid dark red, red, and blue arrows indicate the 1033-nm, 1004-nm, and 408nm lasers. The dotted arrows in the level schemes indicate off-resonant lasers

Before any measurements were performed, starting with the parameters listed in Tab. 2.1 and 2.2, simulations on collinear Raman spectroscopy were performed using two approaches:

1. The analytical solutions derived in Sec. 2.2.1 within the approximations made in Sec. 2.2.
2. The *qspec* python package developed by P. Müller [39], that numerically solves the Liouville equation of the full Hamiltonian, including spontaneous decay and polarization effects, by differentiating between magnetic substates and different laser polarizations.

4.1   Calculating Initial Parameters

Initial parameters for both experiment and simulations were calculated in preparation for the first measurements on the $S_{1/2} \rightarrow D_{3/2}$ Raman transition.

Realistic laser powers P_1 and P_2 and beam waist $w_{1,2}$ were chosen for the 408-nm and 1004-nm laser, to calculate the Rabi frequencies Ω_{nj}. Then for a specified kinetic energy, defined by $U_{\mathrm{acc}} - U_{\mathrm{pump}}$, and from the length of the pump region, the interaction time t_{int} between ions and lasers was calculated. From this, the two-photon Rabi frequency $\Omega_{\mathrm{R,target}}$ required for a π-pulse excitation was calculated, which allows solving Eq. (2.20) for the detuning Δ_{target}. This was used to calculate the laser frequencies $\omega_{\mathrm{L1,2,rest}}$ in the ion rest-frame, δ (including photon recoil contributions), and the AC-Stark shifts Ω_n^{AC}. ω_{L2} was adjusted to compensate for $\delta - \delta_{\mathrm{AC}}$. The laser frequencies were then transformed into the laboratory-frame frequencies $f_{\mathrm{L1,2}}$. Using $f_{\mathrm{L1,2}}$ the voltages required to resonantly probe the

J. Spahn, *Pioneering Raman Transition Techniques in Collinear Laser Spectroscopy for High-Precision Experiments*, BestMasters,
https://doi.org/10.1007/978-3-658-50605-6_4

meta-stable state through the $D_{5/2} \rightarrow P_{3/2}$ or $D_{3/2} \rightarrow P_{3/2}$ transition in the FDR, $U_{\mathrm{FDR,SP}}$ and $U_{\mathrm{FDR,DP}}$, were calculated.

The determined parameters are listed in Tab. 4.1. Since the line shape in Eq. (2.30) is a $\sin^2$-modulated Lorentzian with a FWHM of Ω, the width of the Raman peak can be estimated to be $\Omega/2 \approx 0{,}25\,\mathrm{MHz}$.

In previous attempts described in [17], a detuning $\Delta = 10\,\mathrm{GHz}$ of was chosen, and laser powers of $P_1 = 1\,\mathrm{mW}$ and $P_2 = 7\,\mathrm{mW}$ were used. The dipole moments of calcium [14] are of the same order of magnitude as the ones considered here, and the two-photon Rabi frequency Ω_R is proportional to $\frac{\sqrt{I_1 I_2}}{\Delta}$. Hence, it can be estimated, that ten times the laser power would have been required to achieve efficient population transfer.

Table 4.1 Parameters calculated with the analytical approximation in Sec. 2.2.1 for the $S_{1/2}$-$P_{3/2}$-$D_{3/2}$ Λ-scheme in $^{88}\mathrm{Sr}^+$ using the values in Tab. 2.1 and 2.2. Unless specified otherwise, all simulations performed within this work use the listed laser frequencies, laser intensities, and voltages, as well as the values in Tab. 2.1 and 2.2 as starting parameters

Parameter	Value	Parameter	Value
U_{acc}	$20\,\mathrm{kV}$	U_{pump}	$500\,\mathrm{V}$
w_1	$1\,\mathrm{mm}$	w_2	$1{,}5\,\mathrm{mm}$
P_1	$0{,}1\,\mathrm{mW}$	P_2	$20\,\mathrm{mW}$
t_{int}	$5{,}8\,\mathrm{\mu s}$	$\Omega_{\mathrm{R,target}}$	$542\,\mathrm{kHz}$
I_1	$64\,\mathrm{W/m^2}$	I_2	$5659\,\mathrm{W/m^2}$
Ω_{11}	$31{,}24\,\mathrm{MHz}$	Ω_{22}	$67{,}73\,\mathrm{MHz}$
Ω_{12}	$294{,}56\,\mathrm{MHz}$	Ω_{21}	$7{,}18\,\mathrm{MHz}$
Δ_{target}	$-1953{,}4\,\mathrm{MHz}$	$\Delta_{\mathrm{target}}/2\pi$	$-310{,}9\,\mathrm{MHz}$
$\omega_{\mathrm{L1,rest}}/2\pi$	$734\,989\,513{,}4\,\mathrm{MHz}$	$\omega_{\mathrm{L2,rest}}/2\pi$	$298\,615\,823{,}2\,\mathrm{MHz}$
f_{L1}	$734\,482\,446{,}5\,\mathrm{MHz}$	f_{L2}	$298\,409\,747{,}1\,\mathrm{MHz}$
Ω_1^{AC}	$125\,\mathrm{kHz}$	Ω_2^{AC}	$587\,\mathrm{kHz}$
δ_{AC}	$-462\,\mathrm{kHz}$	$\delta_{\mathrm{AC}} - \delta$	$0{,}1\,\mathrm{kHz}$
Ω_R	$542\,\mathrm{kHz}$	Ω	$542\,\mathrm{kHz}$
$\Omega \cdot t_{\mathrm{int}}/\pi - 1$	$3 \cdot 10^{-8}$	$\frac{\Omega_\mathrm{R}}{\Omega} - 1$	$-3 \cdot 10^{-8}$
$U_{\mathrm{FDR,SP}}$	$476\,\mathrm{V}$	$U_{\mathrm{FDR,DP}}$	$441\,\mathrm{V}$

4.2   Comparing the Two-level Approximation to Full Hamiltonian

4.2.1   Population of the $D_{3/2}$ state

To validate the approximations made in Sec. 2.2, the analytic simulations with the parameters listed in Tab. 4.1 were compared to numeric simulations performed with *qspec*.

Fig. 4.1 a) shows the expected $D_{3/2}$ population $|\rho_{22}|^2$ according to Eq. (2.28), compared to the population calculated by *qspec* for different polarization configurations. While the line shapes are in good agreement, the amplitudes in the *qspec* spectra are lower. The reason for this becomes evident when comparing Rabi frequencies Ω_{nj} calculated by *qspec*, which includes polarization effects, to the value obtained by using the effective dipole moment used in the analytical solution.

For π/π-polarized light ($\Delta m = 0$) the Rabi frequencies are identical in the $S_{1/2}$, $m_J = \pm 1/2$, $P_{3/2}$, $m_{J'} = \pm 1/2$ channel. However, in the $P_{3/2}$, $m_J = \pm 1/2$, $D_{3/2}$, $m_{J'} = \pm 1/2$ channels, the only two channels allowed for linearly polarized light, *qspec* Rabi frequencies are a factor $\sqrt{5}$ smaller than the value obtained using the effective dipole moment due to the inclusion of polarization effects. Thus, only a $\pi/\sqrt{5}$ excitation is achieved and the population transfer is lower.

For isotropically polarized light, the agreement between the Rabi frequencies of the individual channels and the effective value is better. Since the effective dipole approximation is an effective average for isotropically polarized light, the Rabi frequencies in the individual channels do not match the effective average, limiting the population transfer.

In the case of a π/σ^+-configuration, the Rabi frequencies of the allowed $P_{3/2} \rightarrow D_{3/2}$ transition channels calculated by *qspec* are a factor of 1.1 to 1.3 higher than the Rabi frequency calculated using the effective dipole moment. Hence, population transfer increases to 80 %, indicating that optimizing the polarization of the lasers might increase the signal in the experiment.

This discrepancy in Rabi frequencies impacts the laser power required to drive optimal population transfer at a given interaction time. In Fig. 4.1 b) the $D_{3/2}$ population, obtained using *qspec* at different laser powers P_2 of the 1004-nm laser is shown. As expected from the difference of a factor of $\sqrt{5}$ in the Rabi frequency in the case of π/π polarized lasers, a five times higher laser power is required to maximize the population transfer. For isotropic light, population transfer is maximized at the calculated laser power, and for π/σ^+ polarized light maximal population transfer is achieved at about 3/4 of the power calculated for isotropic polarization. Even when optimizing the laser power to match a π pulse, population transfer is

still limited by the dampening of the Rabi oscillation (see Sec. 4.2.2), which is not
included in the analytical solution.

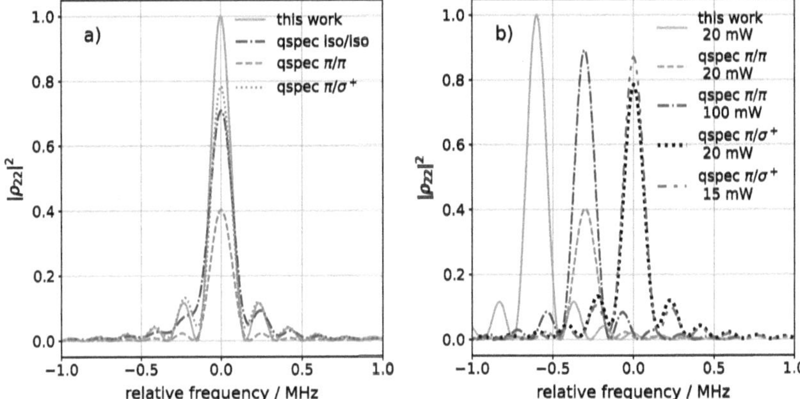

Fig. 4.1 Population of the $D_{3/2}$ state calculated using the analytical approximation (this
work, Eq. 2.28) compared to simulations performed using *qspec*. In a) P_2 is 20 mW in all
spectra, while in b) P_2 values are indicated in the legend. For better comparison, in a) the
peak positions were aligned on the x-axis to compensate for the polarization and laser power
dependent AC-Stark shifts. In b) *qspec* simulations are grouped by polarization configuration,
which is labeled as the polarization of the 408-nm laser/polarization of the 1004-nm laser in
the legend. The relative position at the x-axis is without meaning

4.2.2   Rabi Oscillation

Using the ansatz in Eq. 2.40, the time evolution of the Hamiltonian in Eq. 2.6 and
the time evolution of the full Hamiltonian, simulated using *qspec*, were compared.
The power of the 408-nm laser and both laser frequencies were fixed to the values
in Tab. 4.1.

In Fig. 4.2 a) the state population of all three states is shown, using the parameters
of Tab. 2.1 and 4.1, *i.e.*, a laser power of $P_2 = 20$ mW for the analytical simulation.
Qspec simulations were performed for linear polarization and $P_2 = 100$ mW to
match the Rabi frequency of the analytical solution (Eq. (2.28)) for better comparison. With less than 0.1%, the population of the intermediate $P_{3/2}$ state is negligible.
The time evolution is well described by a Rabi oscillation between the $S_{1/2}$ ground
state and the metastable $D_{3/2}$ state. *Qspec* simulations include spontaneous decay,

leading to a dampening of the Rabi oscillation, which limits the achievable population transfer.

Figure 4.2 b) compares the time evolution obtained using *qspec* with different polarization configurations and laser powers P_2 of the 1004-nm laser. As expected from the analytical solution, changing the laser power leads to a shift in Rabi frequency Ω_R. Depending on the polarization configuration, the different Rabi frequencies of different transition channels lead to a superposition of oscillations and a collapse of the Rabi oscillation, as can be seen in the simulations for isotropic polarization or π/σ^+ polarized lasers. While this limits the population transfer, this also implies that the $D_{3/2}$ state is populated, even if the pulse length exceeds π.

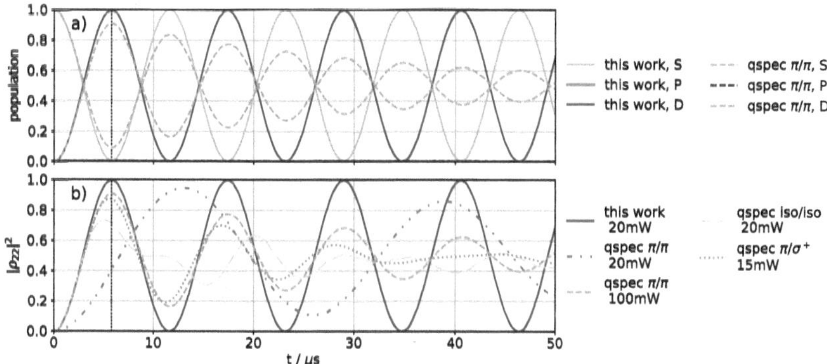

Fig. 4.2 a): Population of the $S_{1/2}$, $P_{3/2}$, and $D_{3/2}$ states calculated using the analytical approach, compared to *qspec* simulations, plotted against t, the interaction time between ions and lasers. For the analytical solution, P_2 is 20 mW, and for the *qspec* simulation linearly polarized light is used and P_2 is 100 mW. b): Population $|\rho_{22}|^2$ of the $D_{3/2}$ state. P_2 and polarizations are listed in the legend. The dotted black vertical line at 5,8 µs indicates the interaction time targeted for this experiment

4.3   Energy Distribution

To study the impact of the ion energy distribution, the population of the $D_{3/2}$ state $|\rho_{22}|^2$ was calculated within the analytical approximation using Eq. (2.28) and Eq. (2.40) for different ion energies, laboratory laser frequencies f_{L1} of the 408-nm laser, and interaction times. The frequency f_{L2} of the 1004-nm laser was fixed to the frequency in Tab. 4.1.

From the measurements shown in Fig. 3.3 and a typical power of 150 W applied to the SIS at COALA, a source temperature of 2150 °C to 2250 °C can be estimated. Assuming a Maxwell-Boltzmann distribution, and considering the ions can have an initial velocity component parallel and antiparallel to the ion beam direction, this yields an energy width of $0,46(1)$ eV, translating to a Doppler width of $3,5(1)$ MHz. This does not include pressure broadening, voltage gradients, or space charge effects inside the ion source.

Realistic line shapes, for the chosen experimental approach described in Sec. 3.4, were obtained by numerical convolution of the line shape in Eq. (2.30) with normally distributed ion energies with a standard deviation ΔE. The interaction time was set to $5,8\,\mu$s. Two examples for a) $\Delta E = 1$ meV and b) $\Delta E = 460$ meV are shown in Fig. 4.3. While for a ΔE of the order of magnitude of the width of the Raman peak, the $\sin^2()$ structure in the peak flanks is still visible, it is averaged out for higher energy widths. For the experimentally expected energy width of 460 meV, the narrow Raman peak acts like a delta function and projects the energy distribution onto the frequency of the scanned laser. Thus, the line shape is, up to a normalization factor, given by the shape of the energy distribution. A linear dependence of the resonant laser frequency from the ion energy can be seen in b). As described in Sec. 2.3.1 the slope of this dependency is given by the difference ΔD in differential Doppler factors. For $^{88}\text{Sr}^+$ at an energy of $19,5$ keV, $D(f_{L1}) = 13.00$ MHz/eV and $D(f_{L2}) = 5.28$ MHz/eV, resulting in $\Delta D = -7.72$ MHz/eV.

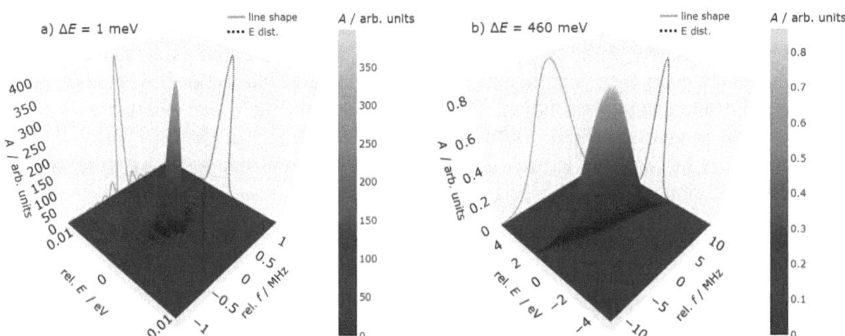

Fig. 4.3 Amplitude A of $D_{3/2}$ state population in arb. units, for an ion beam with normally distributed energies with a standard deviation of 1 meV (a)) and 460 meV (b)). f is the frequency of the 408-nm laser, relative to the frequency f_{L1} in Tab. 4.1 and E is the ion energy relative to 19,5 keV. The black dotted line in the ΔE-A-plane is the energy distribution and the orange line in the f-A-plane is the marginalized line shape. Both are normalized for better visualization

The Rabi oscillation expected, when scanning the 408-nm laser over ions with a velocity distribution, was investigated, by calculating the $D_{3/2}$ population for different ion energies, laser frequencies f_{L1}, and interaction times, using Eq. (2.40). The state populations were weighed with a Gaussian energy distribution from the ion production process, and the total $D_{3/2}$ population for a given frequency and interaction time was calculated by integrating over all corresponding energy contributions. An exemplary heat map of the expected signal for an energy width of 460 meV is shown in Fig. 4.4. Minimal dampening of the $D_{3/2}$ population over time can be seen. Since spontaneous decay was not included in the analytical approach, it can be attributed to the ion energy width. As pointed out in Sec. 2.2.3, a shift in the ion energy shifts Δ and thus Ω_R. The superposition of oscillations with different frequencies leads to a collapse of the Rabi oscillation. This effect increases with the width of the ion energy distribution. It is amplified if contributions from different magnetic substates, and hence different Rabi frequencies, are considered.

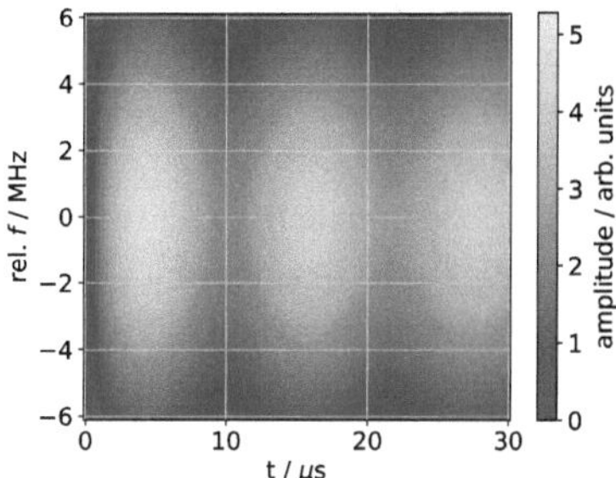

Fig. 4.4 Expected oscillation of $D_{3/2}$ state population, for an ion beam with normally distributed energy with a standard deviation of 460 meV. f is the frequency of the 408-nm laser, relative to the frequency f_{L1} in Tab. 4.1 and t is the interaction time in μs. The heat map indicates the $D_{3/2}$ state population in arb. units

4.4   Simulations of the Double Raman Scheme Including Spatial Intensity Distributions

Simulations on a $S_{1/2} \rightarrow D_{5/2}$ Raman interaction and a sequential $D_{5/2} \rightarrow D_{3/2}$ Raman transition at COALA, as described in 2.4 and 3.5.2 were done using the analytical expression for the time evolution of a four-level system given in Eq. 2.40. Initial parameters were calculated as shown in Sec. 4.1. Like in Sec. 4.3, the ion energy distribution was included to investigate Doppler broadening and the possibility of performing Doppler-free collinear Raman spectroscopy using the double Raman scheme. Additionally, spatial intensity distributions of laser and ion beams were added. This was done by generating a two-dimensional grid in coordinate space representing the beam cross-section. Ion and laser beam intensity are fixed for each grid point and the time evolution of the system is then calculated for the first Raman transition (interaction time $t_{\text{interact},1}$ $=5,8\,\mu s$) at a voltage $U_{\text{pump},1}$ for $t \in [0, t_{\text{interact},1}]$. The final states after the first interaction are taken as initial states to calculate the time evolution during the second interaction at a voltage $U_{\text{pump},2}$ and with an interaction time of $t_{\text{interact},2}$ $=1,4\,\mu s$. Laser and ion beam intensities, as well as the ion energies, were assumed to be constant during the individual interactions. However, the developed code can be easily adapted to include *e.g.*. field penetrations or converging/diverging beam profiles.

Simulations shown in this section were performed for laser beams with Gaussian radial intensity distributions and beam waists of w_{408} $=1\,mm$, w_{1004} $=2\,mm$, and w_{1033} $=1,5\,mm$ for the 408-nm, 1004-nm, and 1033-nm laser beams. The ion beam intensity was assumed to be normally distributed with a 2σ-diameter of $5\,mm$ and the ion energy was assumed to be normally distributed with an average energy of $20\,keV$ and a standard deviation of $500\,meV$. The voltages of the interaction regions were set to $U_{\text{pump},1}=250\,V$ and $U_{\text{pump},2}=150\,V$.

A 3π-pulse was targeted for both interactions and all beams were presumed to be perfectly aligned. Figure 4.5 shows the beam profiles of all three lasers and the ion beam. In Fig. 4.6 the time evolution of the population of $D_{3/2}$, $D_{5/2}$, and $P_{3/2}$ states are shown as fractions of the total ion beam. The total beam population was obtained by marginalizing over all beam energies, weighted by the ion energy distribution, as well as over the two dimensional spatial grid. Due to the assumed large energy width of the ion beam and the narrow Raman transition only a fraction of 0.8 % is transferred to the $D_{5/2}$ state. Additionally, the time evolution in the center of the beam, obtained from the center grid point and scaled by a factor of 100 for better comparison, is shown.

In the beam center, the population behaves as expected: A 3π Raman transition into the $D_{5/2}$ state is followed by a consecutive 3π-pulse from the $D_{5/2}$ state into the $D_{3/2}$ state. All ions transferred into the $D_{5/2}$ state during the first interaction are transferred into the $D_{3/2}$ state in the second interaction. If the population across the entire beam is considered, two things stand out:

1. The Rabi oscillation collapses. This is especially pronounced in the first interaction, where the difference in diameters of the laser beams is largest. Population transfer increases beyond a single π-pulse and does not return to zero. The oscillation frequency of the total beam population is not given by the two-photon Rabi frequencies of the individual Raman transitions.

2. The different spatial intensity distributions of the laser beams limit the population transfer in the second Raman transition.

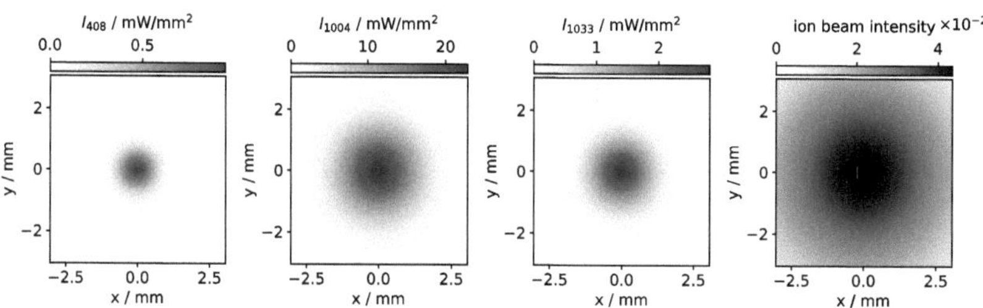

Fig. 4.5 Intensity distributions of the 408-nm, 1004-nm, and 1033-nm laser, as well as the ion beam used to simulate the Double Raman scheme. The lasers have a Gaussian beam profile with a beam waist $w_{408} = 1$ mm, $w_{1004} = 2$ mm, and $w_{1033} = 1,5$ mm, which corresponds to the experimental conditions. I_{408}, I_{1003}, and I_{1033} are the intensities of the three lasers, given in mW/mm^2. The ion beam intensity is normally distributed with a diameter of 5 mm, and is given as a fraction of the total beam per mm^2. x and y are the horizontal and vertical distances from the central beam axis in mm

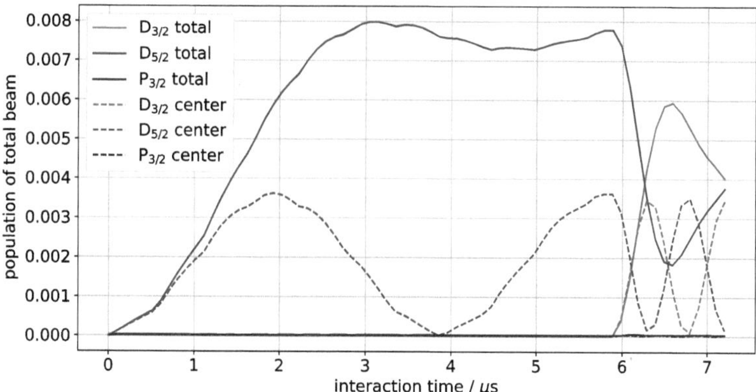

Fig. 4.6 Simulations of the time-dependent population transfer for a $5,8\,\mu s\ 3\pi\ S_{1/2} \rightarrow D_{5/2}$ Raman transition, and a sequential $1,4\,\mu s\ 3\pi\ D_{5/2} \rightarrow D_{3/2}$ Raman transition. The assumed beam profiles are shown in Fig. 4.5. An ion beam energy standard deviation of 500 meV was assumed. Solid lines indicate the population integrated over all beam positions and energies, and dashed lines indicate the time evolution at the center of the beam, integrated over all energies and scaled by a factor of 100 for better visualization

Both can be explained when considering the spatial distribution of the population transfer. Since the laser beam intensities are space-dependent, so is the frequency and the phase of the two-photon Rabi oscillation. Figure 4.7 shows the beam cross-section of the population in the $D_{5/2}$ state after the first interaction ($t = t_{\mathrm{interact},1}$) and of the $D_{3/2}$ population after the second interaction ($t = t_{\mathrm{interact},1} + t_{\mathrm{interact},2}$).

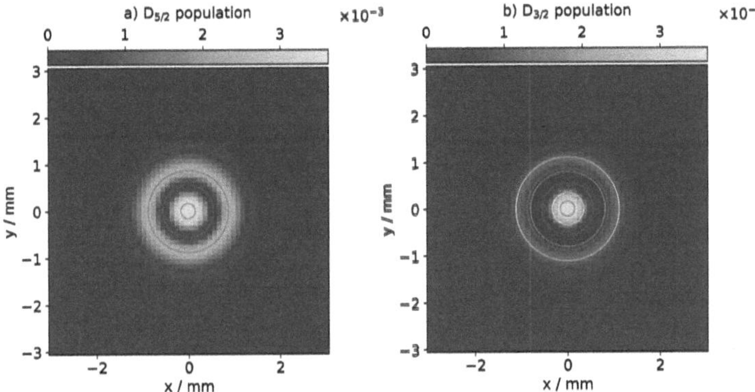

Fig. 4.7 Cross-section of a) the $D_{5/2}$ state population after a 3π pulse $S_{1/2} \rightarrow D_{5/2}$ transition and of b) the $D_{3/2}$ state population after a consecutive 3π pulse $D_{5/2} \rightarrow D_{3/2}$ transition. The color map represents the population as a fraction of the total beam per mm². x and y are the horizontal and vertical distances from the central beam axis in mm.
a): In the beam center a 3π-pulse is transferred and the population of the $D_{5/2}$ state is maximal. As the intensity and hence the two-photon Rabi frequency decreases with distance from the center of the Gaussian beam, this area is followed by rings in which a 2π pulse and a 1π-pulse excitation are driven, resulting in a minimum and a second maximum in population transfer. The maxima/minima are marked as red solid/dashed lines. b): Like in a) the areas, in which a 3π-, 2π-, and 1π-pulse are transferred, are marked as white lines. As the laser beam diameters differ between the 408-nm and 1004-nm laser, the rings do not align with the rings of the first interaction. Efficient population transfer in the second interaction is only possible if ions of a given area are efficiently transferred in both interactions. This limits the total population transfer into the $D_{3/2}$ state

The phase of the first two-photon Rabi oscillation decreases with radius, mirroring the laser beam intensity. While in the center a 3π-pulse leads to maximal population transfer, further out the intensity is only high enough to transmit a 2π-pulse, resulting in no population transfer in that area of the beam. Even further out, the laser intensity only suffices for a 1π-pulse, subsequently, a second maximum in population transfer arises, leading to a ring-like pattern. The emergence of this pattern with the interaction time can be seen in Fig. 4.8.

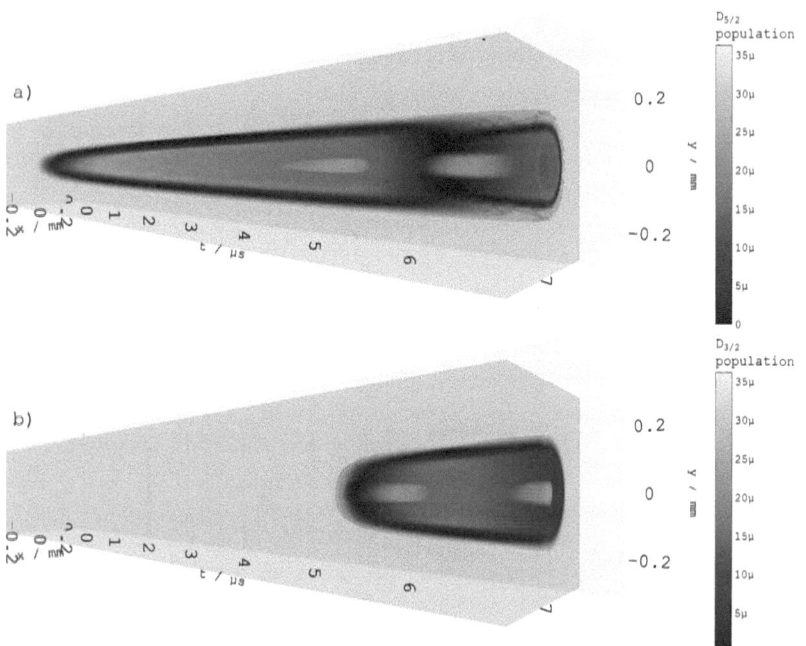

Fig. 4.8 3D render of the space dependent population of a) the $D_{5/2}$ state and b) the $D_{3/2}$ state for a 3π $S_{1/2} \rightarrow D_{5/2}$ transition with $t_{\text{interact},1} = 5,8\,\mu s$ and a consecutive 3π $D_{5/2} \rightarrow D_{3/2}$ Raman transition with $t_{\text{interact},2} = 1,4\,\mu s$. The ion energy is normally distributed with a standard deviation of $500\,\text{meV}$, beam profiles are assumed to be as shown in Fig. 4.5. x and y are the horizontal and vertical distances from the central beam axis in mm, and t is the interaction time in µs.
a) At $t = t_{\text{interact},1}/3 = 1,9\,\mu s$ a 1π-pulse is reached in the first interaction in the center of the beam, which transfers the population from the $S_{1/2}$ to the $D_{5/2}$ state. The laser intensity, and thus the two-photon Rabi frequency, decreases with the distance from the beam center. Therefore, to compensate for the phase increase with t, this 1π-pulse ring moves away from the beam center with increasing time. At $t = t_{\text{interact},1}$ a 3π-pulse is reached at the beam center. b) Here the same happens for the $D_{5/2} \rightarrow D_{3/2}$ transition, except that ions can only be transferred when the areas of population transfer align. Hence, the 1π-pulse ring of the second interaction is only visible when it aligns with the 1π-pulse or 3π-pulse of the first interaction. The same applies to a 2π-pulse back into the $D_{5/2}$ state

This also applies to the second Raman transition, limiting the total population transfer into the $D_{3/2}$ state. Differences in laser beam diameters enhance this effect, as areas of minimal and maximal population transfer are shifted. If ions of a specific area experience a 3π-pulse in the first interaction, but a 2π-pulse in the second interaction, they do not contribute to the total population of the $D_{3/2}$ state. This reduces the signal measured when probing the $D_{3/2}$ state.

Furthermore, as the total population transfer of the beam is given by the integral over the entire beam cross-section, and thus by an integral over oscillations of different frequencies, contributions of different frequencies are averaged out, resulting in the collapse of the Rabi oscillation.

The efficiency of the total population transfer into the $D_{3/2}$ state as well as the value to which the Rabi oscillation collapses highly depend on the beam profiles and laser intensities.

Figure 4.9 shows the population in the $D_{5/2}$ and $D_{3/2}$ state after the second interaction as a function of the voltages $U_{\mathrm{pump},1}$ and $U_{\mathrm{pump},2}$ applied to the interaction regions. When probing the $D_{5/2}$ state, a Lamb dip is visible in the Doppler-broadened Raman resonance of the $S_{1/2} \rightarrow D_{5/2}$ transition. In the $D_{3/2}$ state population, an unbroadened $D_{5/2} \rightarrow D_{3/2}$ Raman peak can be seen. The position of this peak depends on the voltage applied to the first interaction region. If a higher voltage is applied, faster ions are selected in the first interaction and a higher voltage is required in the second interaction region to fulfill the two-photon resonance condition of the Raman transition. It can also be seen, that, instead of operating the first interaction region as a velocity filter at a fixed voltage and scanning the second interaction potential, the second interaction region can be used as a velocity filter when scanning the first interaction potential.

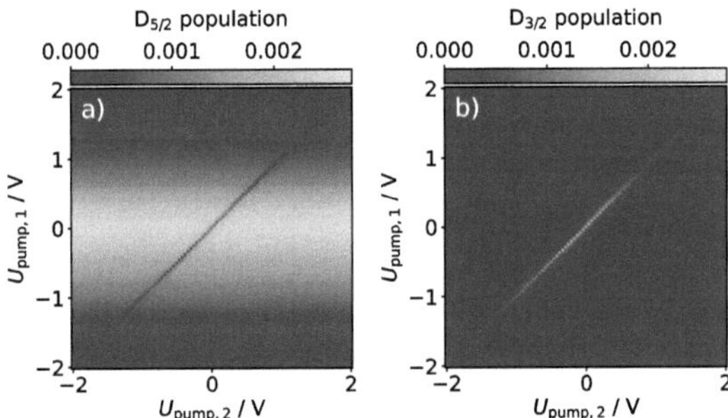

Fig. 4.9 Population of a) the $D_{5/2}$ state and b) the $D_{3/2}$ state after a 3π $S_{1/2} \rightarrow D_{5/2}$ Raman transition, and a consecutive 3π $D_{5/2} \rightarrow D_{3/2}$ Raman transition. $U_{\mathrm{pump},1/2}$ are the voltages applied to the first/second interaction region, relative to 250 V/150 V. The population is shown as a fraction of the total beam, accounting for the ion energy distribution and the spatial intensity distributions of the ion and laser beams

Experimental Results 5

In this chapter the measurements employing single and double Raman schemes are described and the observed Lamb dip and the line shapes are discussed. The Rabi oscillation, respective its collapse, is investigated and compared to the simulations performed in the previous chapter. Furthermore, the impact of the detuning was explored.

5.1 Single Raman Measurements

The first measurements were performed at the $S_{1/2} \to D_{3/2}$ Raman transition using the measurement scheme described in Sect. 3.5.1. An acceleration voltage of $20\,009$ V was used and the optical pumping tube was floated to 200 V. The laser frequencies were calculated as described in Sect. 4.1. For example, for a detuning of -300 MHz, the 408-nm laser was set to 734 479 038 MHz at a typical laser power of 3 mW and the 1004-nm Matisse was set to 298 408 788 MHz at a typical laser power of 350 mW. The AOM blocked the laser for $10,2\,\mu$s and was open for $11,1\,\mu$s for laser-background free measurements. A voltage of 440 V was applied to the FDR to probe the $D_{3/2}$ state.

A time-resolved spectrum of the $S_{1/2} \to D_{3/2}$ Raman transition taken at these settings is shown in 5.1. The remaining background signal is caused by the thermal population of the $D_{3/2}$ state and 408-nm light leaking through the AOM, even when turned off. Due to the finite rise and fall time of the AOM, the suppression of the 408-nm light is slightly worse in the first $2\,\mu$s, leading to an increased amount of detected photons during this time. After the AOM is turned off, ions that did not fully pass the interaction region while the AOM was on, are transferred with a lower probability from the $S_{1/2}$ to the $D_{3/2}$ state, leading to a linear decrease of

© The Author(s), under exclusive license to Springer Fachmedien Wiesbaden GmbH, part of Springer Nature 2026

J. Spahn, *Pioneering Raman Transition Techniques in Collinear Laser Spectroscopy for High-Precision Experiments*, BestMasters,
https://doi.org/10.1007/978-3-658-50605-6_5

fluorescence rate. The peak has a FWHM of 2,07(8) eV and shows a significant substructure, which will be discussed in Sect. 5.3.

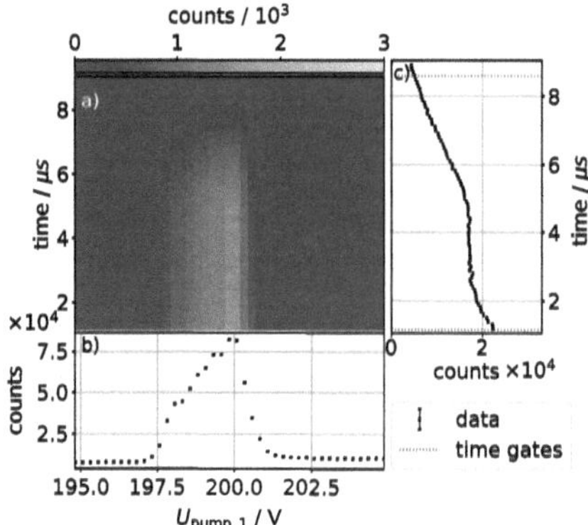

Fig. 5.1 Time-resolved spectrum of the $S_{1/2} \to D_{3/2}$ Raman transition. a): Heat map of the number of measured counts over voltage $U_{\text{pump},1}$ applied to the optical pumping tube (in V) and time t (in μs). The AOM for the 408-nm laser is switched off at $t = 0$ b): Projection of the fluorescence signal on the scan voltage axis. c): Projection of the fluorescence signal on the time axis

The first spectra of the $S_{1/2} \to D_{5/2}$ Raman transition in the single Raman scheme were obtained using the same method and show similar properties.

5.2   Double Raman Measurements

To measure the $D_{5/2} \to D_{3/2}$ Raman transition in the double Raman scheme, an acceleration voltage of 20 009 V was used. The optical pumping tube was floated to $U_{\text{pump},1} = 265$ V to drive the $S_{1/2} \to D_{5/2}$ Raman transition at detuning of -665 MHz. A voltage of $U_{\text{pump},2} = 165$ V was applied to the second pumping region, where the $D_{5/2} \to D_{3/2}$ Raman transition was measured at a detuning of -175 MHz. At the chosen laser frequencies of 734 480 366 MHz for the anticollinear 408-nm laser, 298 408 458 MHz for the anticollinear 1004-nm Matisse laser, and 290 413 085 MHz for the collinear 1033-nm diode laser, the $S_{1/2} \to D_{5/2}$ resonance is at $U_{\text{pump},2} = 390$ V and a detuning of 1020 MHz. Laser powers of $2-3$ mW, 250-300 mW, and $15-20$ mW were used for the 408-nm laser, the 1004-nm Matisse, and the 1033-nm diode laser, respectively. The $D_{5/2}$ and the $D_{3/2}$ states were probed at FDR voltages of 130 V and 190 V.

These settings were determined by aligning the $S_{1/2} \rightarrow D_{5/2}$ and the $S_{1/2} \rightarrow D_{3/2}$ Raman transitions at a voltage of 165 V and a fixed frequency of the 408-nm laser, by adjusting the frequencies of the 1004-nm Matisse laser and the 1033-nm diode laser. The two-photon resonance conditions of the $S_{1/2} \rightarrow D_{5/2}$ and the $S_{1/2} \rightarrow D_{3/2}$ transitions fix the detuning of both the 1004-nm laser and the 1033-nm laser to the detuning of the 408-nm laser, ensuring that the two-photon resonance condition of the $D_{5/2} \rightarrow D_{3/2}$ Raman transition is met at $U_{\mathrm{pump,2}} = 165$ V. The $S_{1/2} \rightarrow D_{5/2}$ and the $S_{1/2} \rightarrow D_{3/2}$ transitions were then shifted in voltage space, by shifting the frequency of the 408-nm laser. This was done to ensure, that at the chosen settings, all Raman peaks are well separated from each other, as required for the double Raman scheme, see Sect. 2.4, as well as from the D2-line, to minimize asymmetry in the background, see Sect. 6.4.4.

Figure 5.2 a) shows a spectrum obtained by scanning the voltage of the second pumping region while driving the $S_{1/2} \rightarrow D_{5/2}$ transition in the optical pumping tube and probing the $D_{3/2} \rightarrow P_{3/2}$ transition in the FDR (see Sect. 3.5.2). Both the $D_{5/2} \rightarrow D_{3/2}$ Raman transition, the $S_{1/2} \rightarrow D_{3/2}$ Raman transition, and the optically pumped D2-line, as well as the $S_{1/2} \rightarrow D_{5/2}$ Raman transition can be seen in the spectrum. Latter is unexpected, since the FDR is set to probe the $D_{3/2}$ state, but can be explained by the high laser powers of the 1004-nm and 1033-nm lasers. This leads to power broadening and prohibits completely separate probing of the $D_{5/2}$ and $D_{3/2}$ state, as the tail of the $D_{3/2} \rightarrow P_{3/2}$ transition, used to probe the $D_{3/2}$ state, overlaps with the $D_{5/2} \rightarrow P_{3/2}$ transition, used to probe the $D_{5/2}$ state, in voltage space. In Fig. 5.2 b) the $D_{5/2}$ was probed, again revealing all four resonances, including the expected dip in the $D_{5/2}$ state population caused by the $D_{5/2} \rightarrow D_{3/2}$ transition.

The simultaneous probing of the $D_{3/2}$ and the $D_{5/2}$ state could be resolved, by decreasing the laser power, at the cost of signal loss in the $D_{5/2} \rightarrow D_{3/2}$ transition, or by further separating the peaks in voltage space. Measuring all three ($S_{1/2} \rightarrow D_{3/2}$, and $S_{1/2} \rightarrow D_{5/2}$, $D_{5/2} \rightarrow D_{3/2}$) Raman transitions represents the opportunity to measure the entire ring closure in the $^{88}\mathrm{Sr}^+$ $^2\mathrm{S}$-$^2\mathrm{D}$ fine structure.

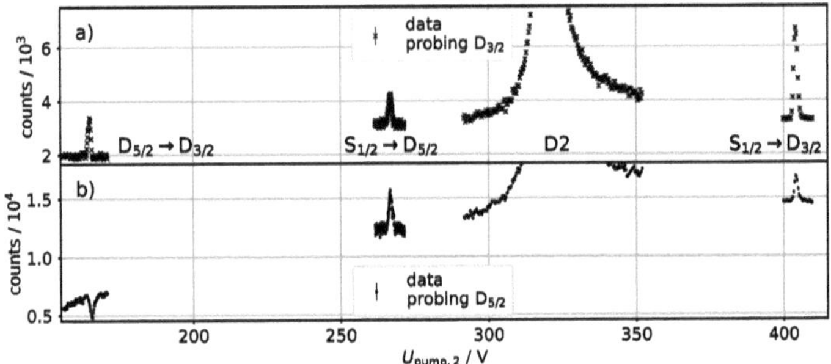

Fig. 5.2 From left to right: Spectrum of the $D_{5/2} \rightarrow D_{3/2}$ and the $S_{1/2} \rightarrow D_{5/2}$ Raman transitions, the optically pumped D2-line, and the $S_{1/2} \rightarrow D_{3/2}$ Raman transition. The voltage at the optical pumping tube is set to match the $S_{1/2} \rightarrow D_{5/2}$ transition resonance and the voltage $U_{\text{pump},2}$ at the second pumping region is scanned. In a) the $D_{3/2}$ state is probed, the $D_{5/2} \rightarrow D_{3/2}$ transition results in a peak. In b) the $D_{5/2} \rightarrow P_{3/2}$ transition is probed, resulting in a dip from the $D_{5/2} \rightarrow D_{3/2}$ transition. Optical pumping of the $D_{5/2}$ and $D_{3/2}$ state via the D2 line and the $D_{3/2} \rightarrow P_{3/2}/D_{5/2} \rightarrow P_{3/2}$ dipole transitions lead to an uneven background signal. The asymmetry in the background signal at the D2-line is caused by laser-ion interactions in the potential gradient at the entrance and exit of the second pumping section. Addressing the $D_{3/2}$ and $D_{5/2}$ states separately is not possible due to power broadening. Driving the $S_{1/2} \rightarrow D_{3/2}/S_{1/2} \rightarrow D_{5/2}$ Raman transitions in the shorter second pumping region, instead of the optical pumping tube, reduces the interaction time and hence the signal height compared to spectra obtained in the single Raman scheme

5.3   Line Shape and Lamb Dip

Comparing the measured line shape of a single Raman transition to the simulations described in Sect. 4.3 shows that the width of the measured peak is one order of magnitude larger than the anticipated width of the unbroadened Raman peak. As discussed in Sect. 2.3, the dominant broadening mechanism is Doppler broadening caused by the velocity width of the ion beam.

Figure 5.3 a) shows the line shapes of the $S_{1/2} \rightarrow D_{5/2}$ transition. Different line shapes were fitted, to described the observed asymmetric shape. An asymmetric Gaussian, with different widths for the left and right sides of the peak center was defined as

$$G(x) = a \cdot \left(e^{-\frac{(x-x_0)^2}{2\sigma^2}} \Theta(x - x_0) + e^{-\frac{(x-x_0)^2}{2(\alpha\sigma)^2}} \Theta(x_0 - x) \right) + b, \qquad (5.1)$$

where a is the amplitude, x_0 is the peak position, b is a constant offset, σ is the standard deviation, α the asymmetry parameter and $\Theta(x)$ is the Heaviside step function. Since the amplitude of the unperturbed Raman transition (Eq. 2.30) is a Lorentzian, and the ion energy is expected to be approximately normally distributed, also an asymmetric Voigt profile was fitted. The Gaussian and Voigt line shapes returned identical fit results with a vanishing Lorentzian contribution for the Voigt profile and a Gaussian width of $\sigma = 4,6(1)$ MHz or $0,61(1)$ eV for the measurement shown in Fig. 5.3. Additionally, a thermal fit, obtained by convolving a velocity distribution of thermal ions with a Lorentzian [49], was performed. However, while yielding realistic temperatures of $2300(300)\,°C$, this fit fails to reproduce the observed peak asymmetry.

Closer investigations showed that the line shape exhibits a triple peak structure, as can be seen in Fig. 5.3. Therefore, a sum of three Gaussian peaks with identical widths ("triple Gaussian") was chosen. A linear term was added to all fit functions to account for the slope in the background, caused by the proximity to the D2-line.

Fitting different line shapes to the $S_{1/2} \rightarrow D_{3/2}$ transition yielded similar results. Widths in the single Raman scheme were between $3,2$ MHz and $5,5$ MHz, corresponding to $0,4$ eV and $0,7$ eV, which is of the order of the energy width expected from the ion source temperature (Sect. 4.3).

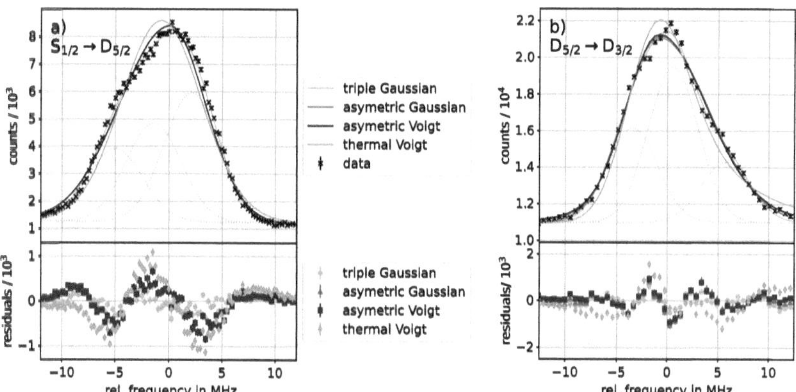

Fig. 5.3 Spectra of a) the $S_{1/2} \rightarrow D_{5/2}$ in the single Raman scheme and b) the $D_{5/2} \rightarrow D_{3/2}$ Raman transition measured in the double Raman scheme by scanning the second pumping region. Top: The black data points indicate the PMT counts and the error bars their statistical uncertainties. The solid lines indicate the fit results and the dotted lines the three peaks of the triple Gaussian line shape. Bottom: The residuals of the different fits. The frequencies on the x-axis are given relative to a) $436\,373\,693$ MHz and b) $8\,405\,350$ MHz

The asymmetry in the background signal and in the D2-line, see Fig. 5.2, was ascribed to laser-ion interactions taking place in the potential gradient at the entrance and exit of the pumping sections. Congruous with this, the direction of the asymmetry changes if a negative or positive voltage is applied.

The peaks of the observed triple peak structure are approximately equidistant with distances of $0,8(1)\,\text{eV}$, and can be explained by different hotspots inside the graphite crucible of the ion source. Since there is a voltage gradient along the crucible, ions produced at different hotspots have different energies. A similar satellite peak was observed at COALA using the same ion source in the past [49]. Another possible explanation could be electron excitations through inelastic collisions in the ion source. The appearance of side peaks and peak asymmetries related to such excitations is a known phenomenon when using charge exchange cells [50].

For all further analysis, a triple Gaussian lineshape was chosen for the $S_{1/2} \to D_{5/2}$ and the $S_{1/2} \to D_{3/2}$ transition in the single Raman scheme, as well as for the $D_{5/2} \to D_{3/2}$ transition in the double Raman scheme. For the $S_{1/2} \to D_{5/2}$ and the $S_{1/2} \to D_{3/2}$ transition in the double Raman scheme an asymmetric Gaussian is used. As the interaction time is shorter than in the single Raman scheme, the population transfer and hence the statistics are reduced. The triple peak structure is not resolved and the fit algorithm does not converge. The asymmetry, the linewidth, as well as the relative peak intensities and the peak positions of the triple Gaussian were fixed to the daily average obtained from unconstrained fits.

As there is no theoretical description of the used line-shape nor is it experimentally fully understood, a systematic uncertainty is introduced for the different measuring schemes, based on differences in transition frequencies obtained using different fit functions.

Contrary to expectation, the $D_{5/2} \to D_{3/2}$ peak width in the double Raman scheme is not in the kHz range, see Fig. 5.3 b). Different line shapes were fitted, resulting in a width of $\sigma = 3,3(1)\,\text{MHz}$ or $0,32(1)\,\text{eV}$. Again, the triple Gaussian fits best to the experimental data.

Several effects could explain the broadening of the $D_{5/2} \to D_{3/2}$ resonance:

1. Relative stability of the laser frequencies on a µs scale: By investigating the beat signals between the different lasers using the spectrum analyzer of the frequency comb this was constrained to be less than $500\,\text{kHz}$.
2. Long-term stability of the diode laser: While the long-term instability of the 1004-nm and 408-nm lasers is below $50\,\text{kHz}$ and negligible, see Sect. 6.4.1, fluctuations of the 1033-nm diode laser frequency cause additional broadening. Based on the measurements performed in Sect. 6.4.1, this can be constrained to $<450\,\text{kHz}$. For the $D_{5/2} \to D_{3/2}$ transition and the Lamb dip in the $S_{1/2} \to D_{5/2}$

transitions in the double Raman scheme, the diode laser drives both sequential transitions. Thus, this effect is, as the total width increases by a factor of $\sqrt{2}$ when convolving two normal distributions of identical widths, enhanced by up to 190 kHz.

3. Instability of the voltage applied to the optical pumping tube: Tests with different high-voltage power supplies used to float the optical pumping tube did not show any improvement in linewidth. From measurements with an ISEG 3 kV high-voltage power supply, which is specified to have a noise of < 100 ppm or 0,3 V, this can be constrained to $<3,1$ MHz.

4. AC-Stark broadening: As described in Sect. 2.3.3, ions interacting with different parts of the laser beam experience different AC-Stark shifts due to the spatial intensity distributions of the laser beam. This might broaden the resonance. However no change in linewidth with the laser intensity was observed, and simulations performed in Sect. 4.4, which include the AC-Stark shift, did not indicate any significant broadening related to this effect.

5. Time of flight broadening: As described in Sect. 2.3.2, a short interaction time with the lasers increases the laser linewidth in the ion rest frame. Using Eq. 2.36 and the interaction time of approximately 1,4 µs, a time of flight broadening of 4 MHz can be estimated. Time of flight broadening has also been observed at COALA before when probing In$^+$ ions in a an interaction region of similar length [51].

The observed broadening of the $D_{5/2} \rightarrow D_{3/2}$ resonance is thus in good agreement with the broadening expected due to voltage fluctuations in the interaction regions and time of flight broadening.

The broadening of the $D_{5/2} \rightarrow D_{3/2}$ transition, and consecutively also of the Lamb dip, implies that the Lamb dip is less pronounced and of similar width as the velocity class transferred to the $D_{5/2}$ state from the ground state in the first interaction, see Fig. 5.4. Hence, fitting the Lamb dip simultaneously to the main peak, when scanning the $S_{1/2} \rightarrow D_{5/2}$ or $S_{1/2} \rightarrow D_{3/2}$ transition in the second interaction, was not possible. Attempts to constrain fit parameters, by subtracting spectra with and without Lamb dip, to isolate the dip line shape, or by fixing the line shape of the main peak, were unsuccessful.

To investigate the Lamb dip, this was overcome by excluding the data points lying on the Lamb dip from the fit, which are depicted in red in Fig. 5.4. A width of 1,0(1) MHz was determined for the dip in the $S_{1/2} \rightarrow D_{3/2}$ Raman transition and a width of 3,3(2) MHz in the $S_{1/2} \rightarrow D_{5/2}$ Raman transition. The ratio of the dip widths approximately corresponds to the differential Doppler factor of $\Delta D = 7.7$ MHz/eV for the $S_{1/2} \rightarrow D_{3/2}$ and $\Delta D = 18.0$ MHz/eV for the $S_{1/2} \rightarrow D_{5/2}$

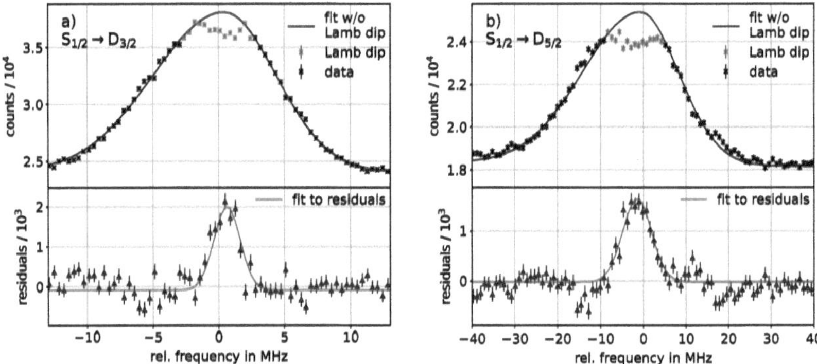

Fig. 5.4 Lamb dip in a) the $S_{1/2} \to D_{3/2}$ and in b) the $S_{1/2} \to D_{5/2}$ Raman transition. In the optical pumping tube, the $S_{1/2} \to D_{5/2}$ transition is driven at a fixed voltage, and the voltage of the second interaction region is scanned over the respective transition, see Sect. 3.5.2. On the x-axis, the frequency is given relative to 436 373 693 MHz and 444 779 043 MHz, respectively. The data points lying on the Lamb dip are indicated in red. The Lamb dip becomes visible as a peak in the residuals, shown in the lower part of the figure. In the lower part a Gaussian was fitted to the residuals. Its width σ is $1,0(1)$ MHz for the $S_{1/2} \to D_{3/2}$ transition and $3,3(2)$ MHz for the $S_{1/2} \to D_{5/2}$ transition

transition. Note that the difference in these values is mostly caused by the different directions of the 1004-nm/1033-nm lasers. This indicates, that the energy selection in the first interaction is limited to $0,15(5)$ eV by either the high-voltage power supply or field penetrations. As ΔD for the $D_{5/2} \to D_{3/2}$ transition is 10.3 MHz/eV, this corresponds to a width of $1,6(5)$ MHz for the $D_{5/2} \to D_{3/2}$ transition, which is half of the observed linewidth. It can be concluded, that the other mechanisms discussed above, cause additional broadening.

As described in Sect. 4.4 and simulated in Fig. 4.9, the position of the Lamb dip depends on the voltage applied to the first pumping section, and impacts the position of the $D_{5/2} \to D_{3/2}$ transition. Figure 5.5 shows this dependence for $U_{\text{pump},1} = 273.2$ V and 272.9 V. Correcting the $U_{\text{pump},1}$ voltage offset using a Doppler factor of $\Delta D = 10.3$ MHz/V, both measurement series yield the same transition frequency.

Due to the complicated fitting procedure and the lack of knowledge of the exact position of the Lamb dip and of possible field penetration into the first pumping section, a systematic uncertainty of 500 kHz is introduced for the $D_{5/2} \to D_{3/2}$ transition frequency.

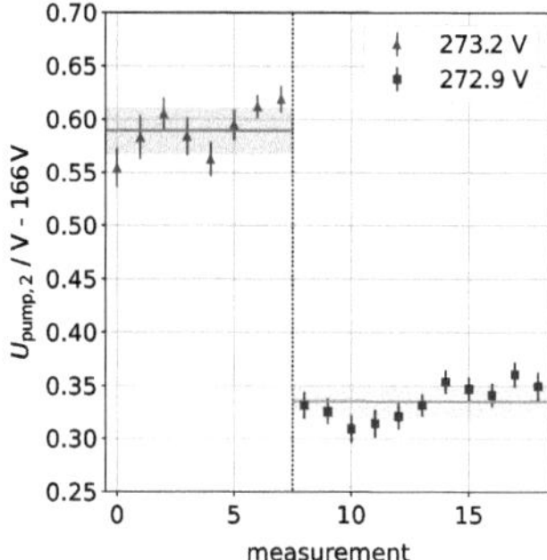

Fig. 5.5 Resonant scan voltages $U_{\mathrm{pump},2}$ for the $D_{5/2} \rightarrow D_{3/2}$ Raman transition at a voltage $U_{\mathrm{pump},1}$ of 273,2 V (red) and 272,9 V (blue) applied to the first optical pumping tube. The solid lines and shaded areas show the average and standard deviation of 166,59(2) V and 166,34(2) V) for 273,2 V and 272,9 V. If the voltage offset in $U_{\mathrm{pump},1}$ is corrected using a Doppler factor of $\Delta D = 10.3$ MHz/V, both measurement series yield the same transition frequency

5.4 Collapse of Rabi Oscillations

To investigate if a Rabi oscillation can be observed, although the simulations performed in Sect. 4.4 (Fig. 4.6) indicate that the spatial intensity distribution of the ion and laser beams might result in a collapse of the Rabi oscillation, the impact of the laser power and the interaction time on the population transfer was investigated. Since the two-photon Rabi-frequency Ω_R is proportional to the square root of the product of both laser intensities $\sqrt{I_1 I_2} \sim \sqrt{P_1 P_2}$, varying the laser power at a fixed interaction time changes the phase of the Rabi oscillation. The same applies for a variation of the interaction time at constant laser powers. This should lead to an oscillation in the amplitude of the signal measured at different laser powers and interaction times.

Signal-to-background (STB) ratios of the $S_{1/2} \rightarrow D_{3/2}$ Raman transition measured at different laser powers are shown in Fig. 5.6 a). The 1004-nm was set to 7 mW, 44 mW, 250 mW and 350 mW while the 408-nm laser power was varied between 10 μW and 3 mW. As the 1004-nm laser is also used to probe the population of the $D_{3/2}$ state, its power also impacts the probing efficiency and, hence, the STB. Comparing the STB ratio at different $\sqrt{P_{1004} \cdot P_{408}}$ shows that the STB becomes independent of P_{1004} for $P_{1004} \geq 44$ mW, as the probing transition is saturated. Hence, all variation in the STB ratio can be attributed to the efficiency of the population transfer of the Raman transition. The STB ratio increases up to

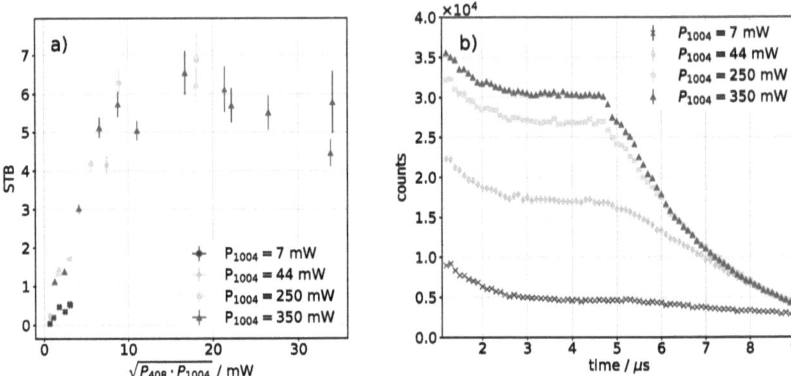

Fig. 5.6 a): Signal-to-background ratio (STB) of the $S_{1/2} \to D_{3/2}$ Raman transition at different laser powers. The STB ratio is plotted over $\sqrt{P_{1004} \cdot P_{408}}$ as this is proportional to the two-photon Rabi frequency.
b): Evolution of the $S_{1/2} \to D_{3/2}$ resonance signal with time for different powers P_{1004} of the 1004-nm laser. The power P_{408} of the 408-nm laser is $1,3(1)$ mW. Shown are the PMT counts over time, starting at the moment the 408-nm laser was turned off with the AOM. The higher but decreasing signal rate below $2\,\mu s$ is caused by scattered laser light during the finite fall time of the AOM. The ions arriving after $5\,\mu s$ have not experienced the full interaction time as the AOM was turned off. No two-photon Rabi oscillation is observed for a changing interaction time. Both the lack of oscillations in the resonance signal over time and the saturation of the resonance signal with increasing laser power is consistent with the collapse of the two-photon Rabi oscillation due to the spatial intensity distribution of the laser and ion beams

$\sqrt{P_{1004} \cdot P_{408}} \approx 15\,\mathrm{mW}$ and saturates. No two-photon Rabi oscillations and no significant decrease can be seen for higher laser powers can be seen, which confirms the assumption of the collapse of the Rabi oscillation due to spatial intensity distributions.

Qualitatively identical observations were made in the $S_{1/2} \to D_{5/2}$ and $D_{5/2} \to D_{3/2}$ Raman transitions.

Figure 5.6 b) shows the signal over time of the $S_{1/2} \to D_{3/2}$ Raman transition at a detuning of 300 MHz measured inside the optical pumping tube at different laser powers P_{1004} of the Matisse at 1004 nm and a constant laser power of $P_{408} = 1,3(1)\,\mathrm{mW}$ of the 408-nm laser. As the 1004-nm laser is also used to probe the population in the $D_{3/2}$ state, the strong increase between $P_{1004} = 7\,\mathrm{mW}$ and 44 mW has to be ascribed to both a more efficient probing and a larger population transfer. As the probing transition is already saturated at 44 mW, the further increase can be chiefly ascribed to a larger population transfer. For the first $5\,\mu s$, the signal is relatively constant as the ions probed in the FDR experienced the full interaction time with the 408-nm laser in the pumping section. The ions arriving after 5 µs have not experienced the full interaction time, as the AOM was turned off while they had not fully passed through the interaction region. This shorter interaction time impacts the phase of the two-photon Rabi oscillation and, under ideal conditions, one would expect two-photon Rabi oscillations. However, even at a laser power of 349(10) mW, for which theoretically an 8π-pulse is transmitted at the center of the beam, no oscillation in the time-resolved signal is visible. This is congruous with the simulation in Sect. 4.4, indicating that the spatial intensity distributions of the laser and ion beams result in a collapse of the two-photon Rabi-oscillation.

5.5 Variation of the Detuning Δ

Changing the detuning Δ influences the two-photon Rabi frequency Ω_R, which is proportional to $\frac{1}{\Delta}$. Hence, a decrease in population transfer is expected, when increasing $|\Delta|$. Measurements were performed from $\Delta = 50\,\text{MHz}$ to $\Delta = 3{,}5\,\text{GHz}$ in the $S_{1/2} \rightarrow D_{3/2}$ transition at a laser power of 350 mW of the 1004-nm laser and 3 mW of the 408-nm laser. The signal-to-background ratios of those measurements are shown in Fig. 5.7 a). Measuring both the Raman peak and the D2-line, as shown in Fig. 5.7 b), allows to determine the detuning with high precision using the distance between both peaks in voltage space and Eq. 2.3.

As expected, a fall in signal proportional to $1/|\Delta|$ was observed. For small detunings below approximately 200 MHz, the $D_{3/2}$ state is already probed inside the optical pumping tube due to an overlap with the 1004-nm transition used to probe the $D_{3/2}$ state. This results in a decrease in the signal-to-background ratio. At detunings smaller than the linewidth of the probing transition, the Raman transition can be driven directly in the FDR, where this is not an issue, and the signal height is not reduced. The $D_{3/2}$ state is then probed on the flank of the probing transition on which the Raman peak is located. This decreases the probing efficiency and thus the measured signal. However, as the background is dominated by the thermal population of the $D_{3/2}$ state, this cancels out in the STB ratio.

With the available laser power, measurements at detunings larger than 3,5 GHz were not possible. Based on the measurements performed within this work, collinear Raman spectroscopy at very large detuning that would be necessary, to investigate isotopes, in which no optical dipole transitions are available, does not seem feasible without significantly increasing the interaction time or laser intensity.

Comparing the height of the optically pumped peak and the Raman peak at a detuning of $\Delta = -558\,\text{MHz}$ in Fig. 5.7 b) shows that Raman transitions can be used

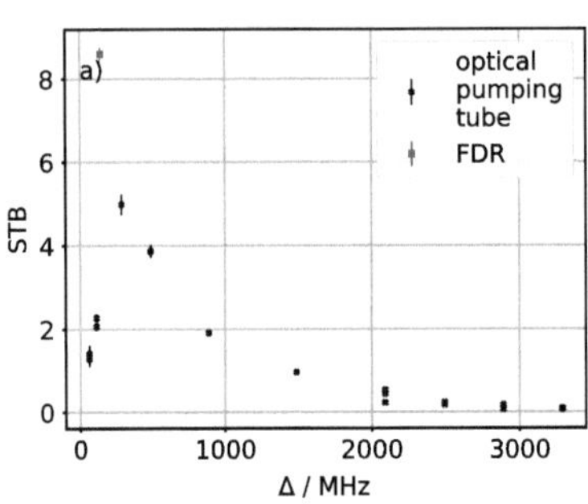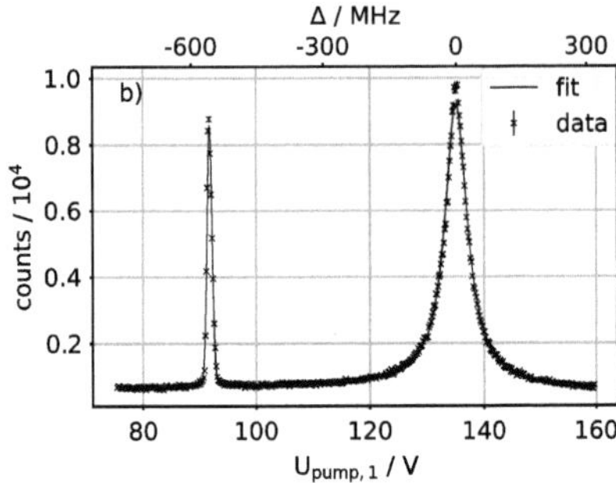

Fig. 5.7 a): Signal-to-background ratio (STB) of measurements of the $S_{1/2} \rightarrow D_{3/2}$ transition performed at different detunings Δ in the single Raman scheme. The data point marked as red square is a measurement in which the Raman transition was driven directly in the FDR. Here, simultaneous probing of the $D_{3/2}$ state at small detunings does not lead to a loss of signal. b): Example spectrum of the $S_{1/2} \rightarrow D_{3/2}$ transition (left peak) and the D2-line (right peak). The distance of both peaks is used to determine Δ. For the measurement shown here the detuning is $-558\,\mathrm{MHz}$

to efficiently transfer the population of an ion beam into the metastable state of a Λ-scheme. This is of particular interest in elements in which the branching ratio into the metastable state is small, and hence, optical population transfer is inefficient.

5.6   Polarization

Qspec simulations indicate that using two linearly polarized lasers with misaligned polarization axes might significantly decrease the efficiency of the population transfer and even lead to a splitting of the unbroadened Raman peak, as described in Sect. 2.2.3. Polarization effects were investigated in the single Raman scheme in the $S_{1/2} \rightarrow D_{5/2}$ and $S_{1/2} \rightarrow D_{3/2}$ transitions by installing a half-wave plate and a quarter-wave plate in the beam path of the 408-nm laser beam. Only minimal variations in signal height for linearly polarized light at different angles were observed. No polarization dependency of the transition frequencies could be resolved. Hence, the topic of polarization will not be further discussed in this work.

Determination of Transition Frequencies

6

6.1 Quasi-Simultaneous col./acol. Measurements

Quasi-simultaneous measurements in collinear/anticollinear geometry were performed on the $S_{1/2} \rightarrow D_{5/2}$ transition by driving a Raman transition at a detuning of $-680\,\text{MHz}$ in the optical pumping tube (see Sect. 3.5.1). Both, the diode laser and the Matisse, were used to produce 1033-nm light, and the collinear and anticollinear resonances were aligned in voltage space by adjusting the collinear and anticollinear 1033-nm laser frequencies $f_{\text{col, 1033}}$ and $f_{\text{acol, 1033}}$ at a constant frequency $f_{\text{acol, 408}}$ of the 408-nm laser. This ensures that the frequency $f_{\text{R, 408}}$ of the 408-nm laser in the ion rest-frame is the same in both directions. Thus, the detuning of the 408-nm laser is identical for the collinear and the anticollinear measurements. Therefore, also the ion rest-frame two-photon resonance conditions are, up to differences in the AC-Stark shift caused by the difference in collinear and anticollinear 1033-nm laser intensities, matched. Hence, in resonance, the frequency of the 1033-nm lasers in the ion rest-frame $f_{\text{R, 1033}}$ is identical for both $f_{\text{col, 1033}}$ and $f_{\text{acol, 1033}}$ and can be calculated using Eq. 2.4. The measurements were performed via Doppler tuning at fixed laser frequencies, and the resonance positions $U_{\text{acol}}/U_{\text{col}}$ determined in voltage space. Taking into account the distance $\Delta U = U_{\text{acol}} - U_{\text{col}}$ between the collinear and anticollinear peak in voltage space, $f_{\text{R, 1033}}$ is given by

$$f_{\text{R, 1033}} = \sqrt{f_{\text{acol, 1033}} \cdot \left(f_{\text{col, 1033}} + \Delta D \Delta U + (\delta f_{\text{AC}}(1033\,\text{acol}) - \delta f_{\text{AC}}(1033\,\text{col}) \right)}, \tag{6.1}$$

where $\Delta D = D(f_{\text{acol, 408}}) - D(f_{\text{col, 1033}})$, is the difference of the differential Doppler factors D, defined in Eq. 2.33 and $\delta f_{\text{AC}}(1033\,\text{col})/\delta f_{\text{AC}}(1033\,\text{acol})$ are the AC-Stark shifts induced by the collinear/anticollinear 1033-nm laser beam.

© The Author(s), under exclusive license to Springer Fachmedien Wiesbaden GmbH,   57
part of Springer Nature 2026
J. Spahn, *Pioneering Raman Transition Techniques in Collinear Laser Spectroscopy for High-Precision Experiments*, BestMasters,
https://doi.org/10.1007/978-3-658-50605-6_6

Since the 408-nm laser interacts with the same ions, in particular with ions of the same velocity, as the 1033-nm laser, the relative Doppler shift $\gamma(1 + \beta) = f_{R,\,1033}/f_{\text{acol},\,1033}$ (Eq. 2.2) can be used to calculate $f_{R,\,408}$ from the anticollinear laboratory frequency $f_{\text{acol},\,408}$,

$$f_{R,\,408} = f_{\text{acol},\,408} \cdot f_{R,\,1033}/f_{\text{acol},\,1033}. \tag{6.2}$$

The $S_{1/2} \rightarrow D_{5/2}$ rest-frame transition frequency is given by the difference of the rest-frame laser frequencies, corrected for the AC-Stark shifts,

$$f_{0,S_{1/2}\rightarrow D_{5/2}} = f_{R,\,408} - f_{R,\,1033} - (\delta f_{\text{AC}}(408) + \delta f_{\text{AC}}(1033)). \tag{6.3}$$

Even though only one of the two lasers is measured in both the collinear and the anticollinear direction, uncertainties in the ion beam energy fully cancel out in this approach.

Anticollinear/collinear and collinear/anticollinear pairs were measured in fast alternation with about two minutes between measurements. Measurements were taken on two different days, at laser powers of $P_{408} = 3\,\text{mW}$ and $P_{1033} = 15/10\,\text{mW}$ in collinear direction and $P_{1033} = 20/45\,\text{mW}$ in anticollinear direction and at a detuning of $-680\,\text{MHz}$. Laser powers of the 1033-nm lasers were chosen such, that the signal intensities matched in collinear and anticollinear directions. Measurements during which either one of the laser frequencies jumped or the electronically measured high-voltage was not stable within 0,03 V were excluded during analysis. The measured $^{88}\text{Sr}\ S_{1/2} \rightarrow D_{5/2}$ and AC-Stark shift corrected (for details see Sect. 6.4.5) transitions are shown in Fig. 6.1. Results are consistent across the two measurement days. The average transition frequency is 444 779 044.18(8) MHz, the weighted average is 444 779 044.21(8) MHz, and the standard deviation of the measurements is 0,64 MHz.

Including systematic uncertainties listed in Tab. 6.1 and discussed in Sect. 6.4, this results in a transition frequency of

$$f_{0,S_{1/2}\rightarrow D_{5/2}} = 444\,779\,044.21(8)_{\text{stat}}(34)_{\text{sys}}\ \text{MHz}.$$

This is in good agreement with the value of 444 779 044.095 485(1) MHz measured by Dubé et al. in an atomic clock [52].

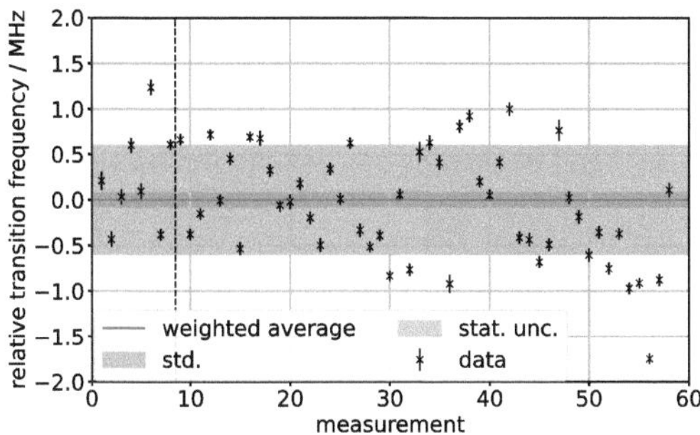

Fig. 6.1 Measured ^{88}Sr $S_{1/2} \rightarrow D_{5/2}$ transition frequencies. The shown values have been corrected for the AC-Stark shift, see Sect. 6.4.5. The black dashed line separates the data taken on the first and second day of measurements. The transition frequencies are given relative to the weighted average of 444 779 044.21 MHz, which is indicated by the solid line. The standard deviation of 0,64 MHz and the resulting statistical uncertainty of 0,08 MHz are marked as shaded areas

Table 6.1 List of uncertainties for the $S_{1/2} \rightarrow D_{5/2}$ collinear/anticollinear measurements. The systematic uncertainty introduced for the peak line shape is based on the difference in absolute transition frequencies obtained using an asymmetric Gaussian and the triple Gaussian line shape. For more details see Sect. 6.4

uncertainty	stat.	AC-Stark	laser	beam overlap	line shape	total
value / MHz	0.08	0.20	0.23	0.01	0.15	0.35

6.2   Double Raman

To extract transition frequencies from measurements in the double Raman scheme, all three Raman transitions and the D2-line are measured, as shown in Fig. 5.2 a). From the known D2 rest-frame frequency and the resonant laser frequency, which is measured with the frequency comb, the kinetic beam energy that matches the resonance condition is extracted (see Sect. 6.4.4). This kinetic beam energy is compared to the electronically measured acceleration potential, and the offset is assumed to be constant for all four transitions within the scan. Using Eq. 2.3 the laser frequencies measured in the laboratory frame are transformed into the ion rest-frame $f_{R,i}$. For each Raman transition, the difference of the laser frequencies in the ion rest-frame $f_{R,1} - f_{R,2}$ is taken. For the $S_{1/2} \rightarrow D_{3/2}$ transition, this is the difference between the 408-nm and the 1004-nm laser, for the $S_{1/2} \rightarrow D_{5/2}$

transition the difference between the 408-nm and the 1033-nm laser, and for the $D_{5/2} \rightarrow D_{3/2}$ transition the difference between the 1004-nm and the 1033-nm laser. The line shapes described in Sect. 5.3 are then fitted. From the two-photon resonance $\delta - \delta_{AC} = 0$ and neglecting any photon momentum transfer, it follows, that

$$(f_{R,1} - f_{R,2}) = (f_{13} - f_{23}) + \delta_{AC} \tag{6.4}$$

$$= f_{12} + \delta_{AC}. \tag{6.5}$$

Thus, the rest-frame transition frequency is obtained from the fitted peak position, corrected for the AC-Stark shift of both lasers. The latter was done as described in Sect. 6.4.5.

Measurements were performed on four different days and are shown in Fig. 6.2. Offsets between the daily averages were observed, in particular, in the $D_{5/2} \rightarrow D_{3/2}$ transition. This is most likely due to different velocity classes being addressed, depending on the position of the Lamb dip, as discussed in Sect. 5.3. Therefore,

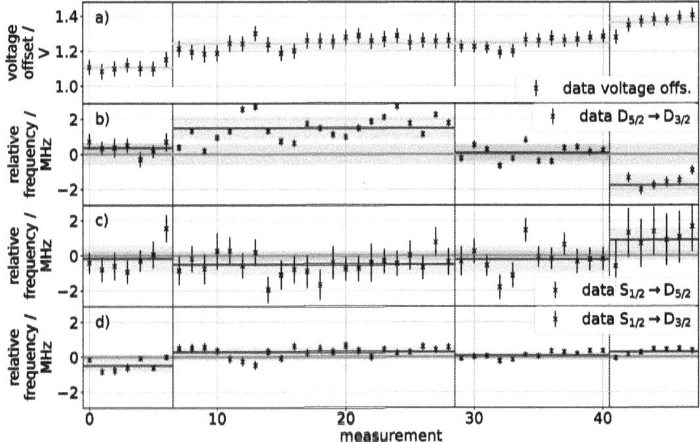

Fig. 6.2 a): Voltage offsets determined from the laser spectroscopic voltage calibration. b)-d): The transition frequency measurements performed in the double Raman scheme. Measurements from different days are separated by vertical dotted lines. All transition frequencies are given relative to the values listed in Tab. 6.2. In a) the solid lines and shaded areas indicate the daily averages of the voltage measurements and their standard deviations. For the transition frequency measurements, the daily weighted averages and standard deviations are marked in blue, and the weighted average and standard deviation of the daily averages are shown in red

the weighted averages and the statistical uncertainties of the daily averages will be given as results.

The AC-stark shift corrected transition frequencies and their uncertainties are listed in Tab. 6.2. Main contributors to the total uncertainty are the uncertainty in the voltage calibration and, for the $D_{5/2} \to D_{3/2}$ transition, the AC-Stark shift and the position of the Lamb dip. Since this transition was measured at a small detuning, the AC-Stark shift is enhanced. The systematic uncertainty of the $S_{1/2} \to D_{3/2}$ transition is significantly smaller, as it was measured at a larger detuning, has a smaller ΔD, and was measured with the two lasers locked to the frequency comb.

The value obtained for the $S_{1/2} \to D_{5/2}$ transition is in good agreement with the frequency measured in quasi-simultaneous collinear/anticollinear measurements, confirming the validity of this method. Relying purely on the electronically measured acceleration voltage would have resulted in an offset of more than 20 MHz, highlighting the importance of performing a voltage calibration against a reference transition if the measurement cannot be performed in both collinear and anticollinear directions.

Table 6.2 List of absolute transition frequencies and their different systematic uncertainties for the measurements performed using the double Raman scheme. The uncertainty listed under line shape includes the frequency shift related to the position of the Lamb dip discussed in Sect. 5.3

transition	frequency [MHz]	stat. [MHz]	AC-Stark [MHz]	voltage [MHz]	laser [MHz]	beam overlap [MHz]	line shape [MHz]	total [MHz]
$D_{5/2} \to D_{3/2}$	8405351.3	0.6	0.8	0.5	0.3	< 0.1	0.5	1.3
$S_{1/2} \to D_{5/2}$	444779043.3	0.3	0.2	0.9	0.3	< 0.1	0.2	1.0
$S_{1/2} \to D_{3/2}$	436373693.6	0.2	0.2	0.5	< 0.1	< 0.1	0.2	0.6

6.3 Ring Closure

Energy conservation dictates that

$$f_{S_{1/2} \to D_{3/2}} + f_{D_{3/2} \to D_{5/2}} - f_{S_{1/2} \to D_{5/2}} = 0. \tag{6.6}$$

For the transition frequencies determined in Sect. 6.2, evaluating this ring closure yields a value of 1.6(1.6) MHz. The ring closure is just fulfilled within 1σ and the relatively large uncertainty is limited by the systematic uncertainties of the $S_{1/2} \rightarrow D_{5/2}$ (main contribution: voltage calibration) and $D_{5/2} \rightarrow D_{3/2}$ (main contribution: AC Stark shift) transition frequencies measured in the double Raman scheme. Alternatively to using the $S_{1/2} \rightarrow D_{5/2}$ transition frequency obtained in the double Raman scheme, the three times more precise result from the collinear/anticollinear measurements (see Sect. 6.1) can be used, yielding a ring closure of 0.7(1.4) MHz.

As the ring closure is mandated by energy conservation, the results obtained in Sect. 6.2 were constrained via Monte Carlo sampling by constraining the ring closure to zero. Direct sampling from the transition frequency measurements in Sect. 6.2 would limit the number of samples, hence, samples were numerically drawn from a normal distribution with averages and standard deviations corresponding to the transition frequencies and their total uncertainties listed in Tab. 6.2. Samples were rejected, if

1. $f_{S_{1/2} \rightarrow D_{5/2}}$ deviates from the value obtained in Sect. 6.1 by more than 2σ or if
2. the ring closure is violated by more than 1 kHz. This limit was chosen, accounting for uncertainties in the kinetic beam energy that might affect the ring closure. For the parameters and laser frequencies used, uncertainties in the acceleration voltage U_{acc} only affect the ring closure by 0, 1 kHz/V. Considering the relative distances of the individual peaks in voltage space, linear field penetrations only affect this ring closure by 5, 3 MHz/V of field penetration per V applied to the interaction region. Based on the measurements shown in Fig. 6.8, the latter can be constrained to be less than 0.01%. The effect of linear field penetration is thus less than 0, 6 kHz.

Out of 15 million drawn samples, 6000 samples were accepted. Figure 6.3 shows the histograms of the accepted samples for each transition. Their average and standard deviation were taken and are listed in Tab. 6.3 and compared to literature values. By construction the ring closure now evaluates to zero, with an uncertainty of 1.1 MHz. The uncertainty of $f_{D_{3/2} \rightarrow D_{5/2}}$ is halved and the result now coincides with the precision measurement in Ref. [52]. The agreement with preliminary results from recent measurements on dipole transitions performed at COALA is reasonable.

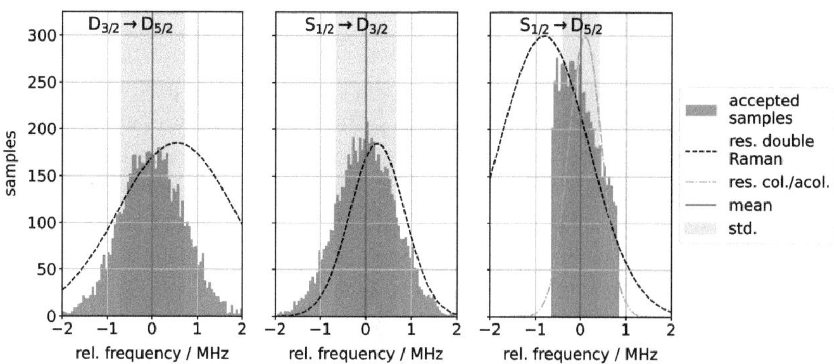

Fig. 6.3 Histograms of the accepted samples for $f_{D_{3/2}\to D_{5/2}}$, $f_{S_{1/2}\to D_{3/2}}$, and $f_{S_{1/2}\to D_{5/2}}$. Samples were drawn from normal distributions with expectation values and standard deviations corresponding to the transition frequencies measured in the Double Raman scheme (Tab. 6.2), indicated by the dotted lines. Likewise, in te right panel, the dash-dotted line indicates the result of the col./acol. measurements in Sect. 6.1. Monte Carlo sampling was used to constrain the obtained transition frequencies by restricting the ring closure to ± 1 kHz and checking for consistency with the $S_{1/2} \to D_{5/2}$ transition frequency determined in the col./acol. measurements within 2σ. The frequencies are given relative to the averages of the accepted samples, marked by the solid lines and listed in Tab. 6.3. The shaded areas mark the 1σ-interval of the accepted samples

Table 6.3 Comparison of the absolute transition frequencies determined within this work using the double Raman scheme, quasi-simultaneous collinear/anticollinear measurements, and by constraining the transition frequencies to the ring closure in the ^{88}Sr$^+$ $S_{1/2}$-$D_{3/2}$-$D_{5/2}$ level scheme. The listed full uncertainties contain statistic and systematic contributions. The COALA values are based on measurements of the D2-line from [33] and recent measurements of $D_{3/2} \to P_{3/2}$ an $D_{5/2} \to P_{3/2}$ transitions that are yet to be published. The uncertainties of the COALA values are preliminary

transition	$f_{D_{3/2}\to D_{5/2}}$ / MHz	$f_{S_{1/2}\to D_{5/2}}$ / MHz	$f_{S_{1/2}\to D_{3/2}}$ / MHz
ring closure	8405350.8(7)	444779044.1(4)	436373693.4(6)
double Raman	8405351.3(1.3)	444779043.3(1.0)	436373693.6(6)
col./acol	/	444779044.21(35)	/
[52]	/	444779044.10(0)	/
COALA	8405349.46(24)	444779043.83(24)	436373694.40(24)

6.4 Systematic Uncertainties

Several factors can affect the transition frequency measurements performed within this work. In the following section, the main systematic uncertainties are discussed and quantified. Namely, the laser stability of the employed laser systems, the correction of the high voltage for field penetrations, contact potentials, and space charge effects in the ion source, and the AC-Stark shift are reviewed. Uncertainties related to the fitted line shape and the shift of the $D_{5/2} \rightarrow D_{3/2}$ transition depending on the Lamb dip position were already discussed in Sect. 5.3.

The total systematic uncertainty σ_{sys} is given by the square root of the sum of squares of the individual contributions $\sigma_{sys, i}$,

$$\sigma_{sys} = \sqrt{\sum_i \sigma_{sys,i}^2}.$$

Measurements were performed at different transitions using two different measuring schemes. Therefore, the systematic uncertainties are evaluated individually for each transition and measuring scheme. Statistic uncertainties are determined separately and added to the total uncertainty $\sigma_{tot}^2 = \sigma_{sys}^2 + \sigma_{stat}^2$.

6.4.1 Laser Frequency

Previous measurement campaigns at COALA have shown that, by locking the Matisse laser to a frequency comb, sub-30 kHz long-term stability can be achieved [33]. To confirm this and investigate the long-term stability of the DL Pro diode laser, a long-term measurement of the comb beat signal was performed for both lasers. The results of both measurements are shown in Fig. 6.4.

For the Matisse laser, the beat signal is stabilized to 60,000 MHz, and a beat of 60,001(13) MHz was measured. No frequency drift can be seen, confirming the stability of the laser.

The diode laser was stabilized to the WSU30 wavemeter, and the frequency set point was adjusted such that the beat signal was at approximately 60 MHz. Like for all measurements within this work, the WSU30 wavemeter was set to auto-calibrate to a HeNe laser every three minutes and read out the laser frequency every 10 ms. A beat signal of 60,94(41) MHz was measured. Furthermore, it can be seen that the stability of the laser improves toward the end of the measurement ($\sigma = 176$ kHz for the last 30 minutes of the measurement). As the laser was already running and locked for several hours by the time the measurement was taken, this cannot be attributed to warm-up effects. Inconsistencies in the stability of the diode laser were observed throughout the entire measurement campaign. They occur if the optical grating used for coarse tuning of diode laser frequency is at the edge of a mode. Therefore, the optical grating was retuned regularly to optimize the stability.

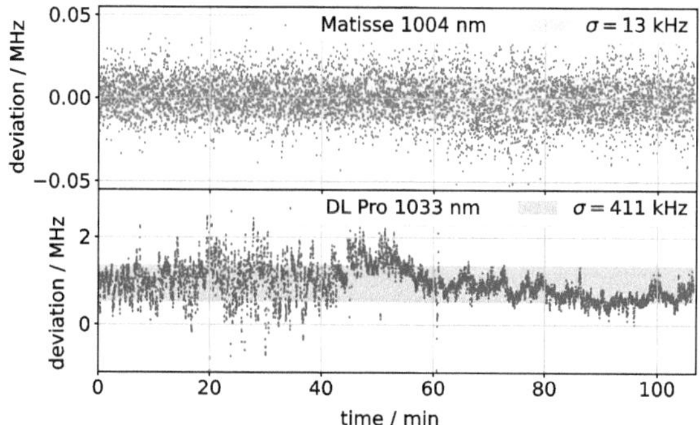

Fig. 6.4 Long-term frequency stability of the used laser systems. The beat deviation from 60 MHz is plotted over the measurement time (in minutes). The upper graph displays the data for the Matisse Ti:Sa laser at 1004 nm. In the bottom graph. the data for the DL Pro 1033-nm laser is shown. The shaded bands indicate the standard deviations. The stability of the diode laser significantly increases after 60 min. This is believed to be caused by the laser drifting toward the center of its optical grating mode

During the spectroscopy measurements, the diode laser was not measured with the frequency comb. Instead, the frequency of the diode laser was monitored and logged using a second wavemeter (WS8-2), allowing to account for deviations from the frequency set point. As explained in Sect. 3.1, the wavemeter can show a systematic deviation from the real laser frequency. For this purpose, the diode laser frequency was measured simultaneously, with the frequency comb for a period of 4−5 min several times a day. From these calibration measurements, a mean offset was calculated and then linearly interpolated for every spectroscopic measurement. Exemplary calibration measurements taken over 9 h can be found in Fig. 6.5. Based on the width of the confidence interval of the fits used to interpolate the wavemeter offset, a systematic uncertainty of 300 kHz is introduced.

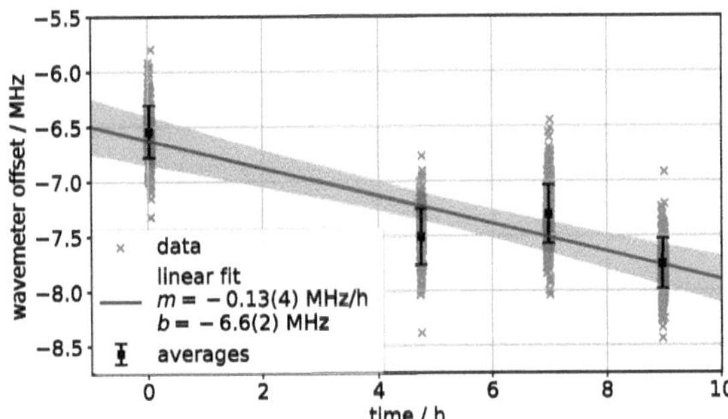

Fig. 6.5 Frequency offset of the WS8-2 used to monitor the diode laser, which is determined from a simultaneous measurement with a frequency comb. The x-axis shows the time t in hours, and the y-axis shows the difference $\Delta f = f_{comb} - f_{WS8-2}$ between the frequency f_{comb} measured using the frequency comb and the frequency f_{WS8-2} measured by the WS8-2 wavemeter. The error bars indicate the average offset and its standard deviation calculated for each measurement. A linear function $\Delta f = mt + b$ is then fitted to those averages (red line). The shaded area marks the confidence interval of the fit

6.4.2 Laser Beam Profiles

Since the AC-Stark shift is intensity-dependent, the beam profiles of all three lasers were measured at different positions using a Thorlabs BC106-VIS beam profiler. The extracted beam diameters are plotted in Fig. 6.6. The beam waists lie outside the laser-ion-interaction regions. Hence, a linear fit was performed, from which the beam diameters inside the interaction regions were extracted. The extracted beam diameters are listed in Tab. 6.4. To account for asymmetries in the beam profile of the diode laser, shown in Fig. 6.7, the geometric mean of the diameters in horizontal and vertical direction was chosen as the effective diameter. This allows comparing laser intensities of different beams at different positions, which is necessary to to correct the measured transition frequencies for the AC-Stark shift, see Sect. 6.4.5.

6.4.3 Beam Alignment

As the Doppler shift depends on the angle between the laser and the ion beam, a potential misalignment of the three laser beams and the ion beam has to be considered. Both the second Matisse and the diode Laser were superimposed with the 408-nm beam at the fiber-coupler of the collinear diode laser and at the dichroic mirror in the

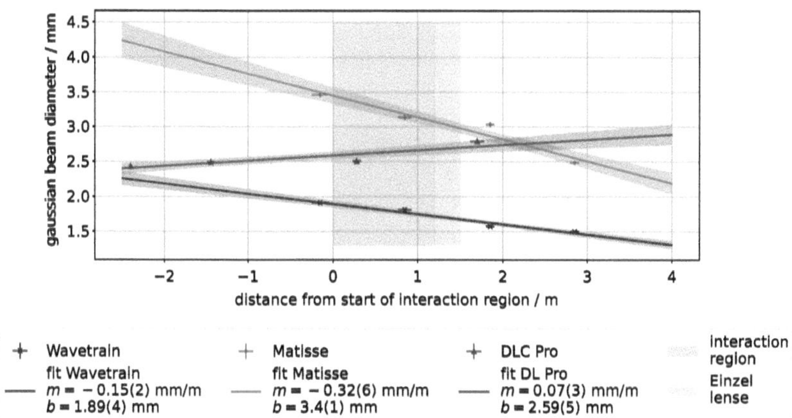

Fig. 6.6 Beam diameter of the laser beams from the Matisse Ti:Sa laser (1004 nm & 1033 nm), the DL Pro diode laser (1033 nm), and the Wavetrain frequency doubling stage (408 nm). The shown value is the geometric mean of the diameters in horizontal and vertical direction, which was determined to compare intensities between different laser beams and to account for asymmetries in the beam profile. The beam diameters are plotted in mm against the distance from the interaction region in m, measured in the direction of the ion beam

laser laboratory (see Fig. 3.1), which have a distance of 10 m. Under the assumption that the lasers are not misaligned by more than half a beam diameter of the 408-nm laser, approximately 1 mm, this results in a maximum angle between the beams of $\alpha_{\mathrm{L}} = \pm 0,20$ mrad. The ion beam and the 408-nm laser beam were aligned using the cameras in the second and third diagnostic stations of the beam line. The two stations have a distance of 2,5 m. As this gives a high-resolution image of the beam

Table 6.4 Laser beam diameters (4σ) extracted from the measurements shown in Fig. 6.6. The uncertainties are estimated based on the variation of the beam diameters within the interaction region as well as the width of the confidence band of the fit in Fig. 6.6. For the Matisse the beam diameter was assumed to be the same for both 1004 nm and 1033 nm

Laser beam	Optical pumping tube	Einzel lens
Matisse	3,3(3) mm	3,0(2) mm
DL Pro	2,6(2) mm	2,7(1) mm
Wavetrain	1,8(1) mm	1,7(1) mm

profiles, a misalignment by less than 0,5 mm can be assumed on each diagnostic station. Hence, a maximum angle of $\alpha_I = \pm 0,4$ mrad is assumed.

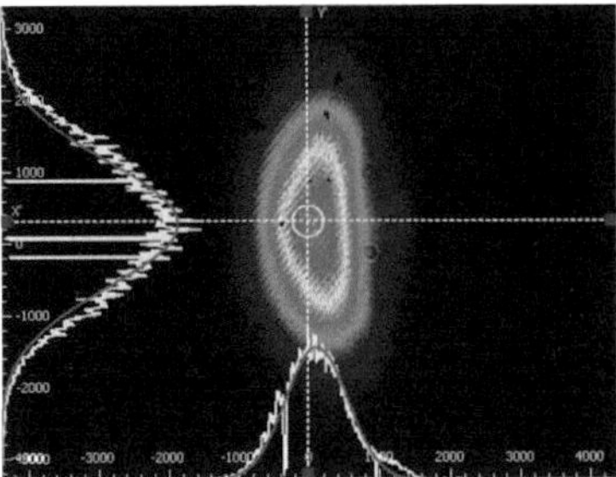

Fig. 6.7 The beam profile of the diode laser measured 30 cm in collinear beam direction from the beginning of the optical pumping tube. Horizontal/vertical distances from the center of the CCD camera are specified in μm on the x-/y-axis. The yellow lines show projections of the laser intensity distribution in horizontal and vertical direction and Gaussian fits to those projections are included in red

For the double Raman scheme, the relevant quantity is the difference of the laser frequencies in the ion rest frame. Using Eq. 2.1 and the angles listed above, the shifts in the extracted rest-frame transition frequencies can be calculated. The largest possible shift occurs if the ion beam and the collinear diode laser are misaligned from the anticollinear 408-nm beam in the opposite direction. For the $S_{1/2} \rightarrow D_{5/2}$ transition frequency measured in the double Raman scheme, this would result in a shift of 77 kHz. However, as the alignment of all beams was optimized daily, this scenario is unlikely. Therefore, to estimate the systematic uncertainty, beam angles were sampled from a uniform distribution within the bounds given above, and the average absolute frequency deviation was calculated. The obtained values are 13 kHz, 24 kHz, and 14 kHz for the $S_{1/2} \rightarrow D_{3/2}$, $S_{1/2} \rightarrow D_{5/2}$, and $D_{3/2} \rightarrow D_{5/2}$ transition.

For quasi-simultaneous collinear/anticollinear measurements, the angle-dependent frequency shift is given by

$$\delta f_0 = \sqrt{f_{\text{acol}} f_{\text{col}}} \left(\gamma \sqrt{(1 + \beta \cos \alpha_{\text{acol}})(1 - \beta \cos \alpha_{\text{col}})} - 1 \right), \qquad (6.7)$$

where $\alpha_{\text{col/acol}}$ are the angles of the collinear/anticollinear laser beams relative to the ion beam, and $f_{\text{col/acol}}$ are the collinear/anticollinear laboratory transition frequencies. This equation is used to calculate the shift of the 1033-nm rest-frame frequency and this shift is then propagated to the 408-nm rest-frame frequency and the $S_{1/2} \rightarrow D_{5/2}$ transition frequency. Again, angles were sampled as described above, and the average absolute frequency shift was calculated, yielding a systematic uncertainty of 13 kHz.

6.4.4 Voltage Calibration

Uncertainties in the acceleration voltage cancel out in quasi-simultaneous collinear/anticollinear measurements. However, both the $S_{1/2} \rightarrow D_{3/2}$ and the $D_{5/2} \rightarrow D_{3/2}$ Raman transitions can not be measured in quasi-simultaneous collinear/anticollinear measurements with the available laser systems. Extracting rest frame transition frequencies of those transitions therefore requires an exact knowledge of the acceleration voltage. Therefore, the amplification factors of the Kepco voltage multipliers were electronically measured by scanning the input voltage between -10 V and 10 V while measuring the amplified voltage using a Keysight 3458 A $8\frac{1}{2}$ digit multimeter. Amplification factors of 50.2907(1) and 50.482(1) were measured for the Kepco 1 and Kepco 2 voltage amplifiers used to scan the optical pumping tube (Kepco 1, single Raman) or the the second interaction region (Kepco 1, double Raman) and to float the optical pumping tube (Kepco 2, double Raman). The uncertainty in transition frequencies resulting from the uncertainty in the amplification factors is below 10 kHz and negligible.

To correct for field penetrations and contact potentials, a laser spectroscopic voltage calibration was performed by measuring the ^{88}Sr D2-line at 408 nm in the second pumping region at different Kepco input voltages. Since this rest frame transition frequency is known at high precision [33], the frequency of the 408 nm-laser can be used to extract the effective acceleration potential using Eq. 2.3. This voltage is then compared to the voltage calculated from the measured acceleration potential U_{acc} applied to the source, the peak position of the D2-resonance, and the electronically determined Kepco amplification factor. Figure 6.8 shows the extracted voltage offsets at different Kepco input voltages. No dependency of the offset from the voltage applied to the second pumping region was found, indicating that the voltage offset is predominantly caused by contact potentials or effects in the ion source (space charge, voltage gradient at the crucible).

Measurements of the voltage offset on different days showed, that the offset is not constant over time. Real-time optical voltage calibrations were therefore performed during the measurements in the double Raman scheme by including the D2 line in the measurements, see Sect. 6.2. Based on the spread of the measured voltage offsets, a systematic uncertainty of 0,05 V is introduced.

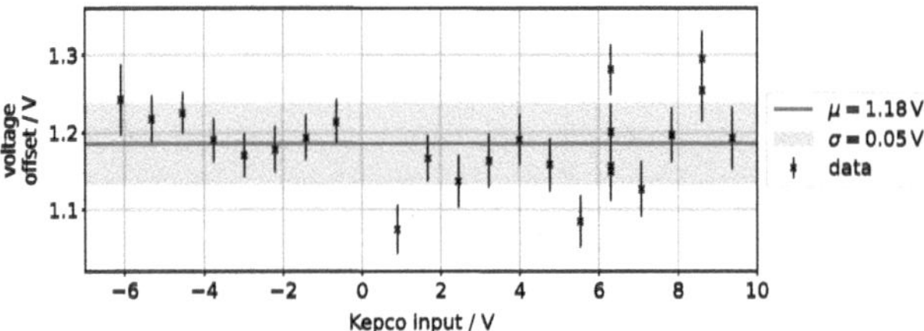

Fig. 6.8 Offset between laser spectroscopically extracted acceleration voltage and electronically measured voltage for different Kepco input voltages. Fit uncertainties of the measurements are indicated by error bars. The solid line and the shaded area indicate the average μ and the standard deviation σ of the measured offset. The measurements were performed at a laser power of $20\,\mu$W

6.4.5   AC-Stark Shift

The AC-Stark shift is negligible when performing collinear laser spectroscopy on dipole transitions at laser powers of up to a few to tens of μW, but at the laser intensities required to drive a Raman transition, the AC-Stark shifts in the kHz to low MHz range are expected. Measurements of all three Raman transitions in the ^{88}Sr$^+$ $S_{1/2}$-$P_{3/2}$-$D_{3/2}$ and $S_{1/2}$-$P_{3/2}$-$D_{5/2}$ Λ-schemes were performed at different laser powers of all three lasers, to measure the AC-Stark shift and correct the measured transition frequencies for this shift. No frequency comb was used for these measurements. Hence, no absolute transition frequencies, only the intensity-dependent relative frequency shift, can be extracted.

Figure 6.9 shows the measured transition frequencies for the $S_{1/2} \rightarrow D_{5/2}$ transition, plotted over the laser power of the Matisse at 1033 nm. The measurements were performed at a detuning of $-483(1)$ MHz in the optical pumping tube.

As expected for the off-resonant AC-stark shift (Eq. 2.19), a linear dependence between the transition frequency and the laser power can be seen. From the slope of a linear fit to the experimental data, an AC-Stark shift of $28(3)$ kHz/mW in the 1033-nm $D_{5/2} \rightarrow P_{3/2}$ transition at the used settings was determined.

Measurements of the AC-Stark shift in the $S_{1/2} \rightarrow P_{3/2}$ transition were performed on the $S_{1/2} \rightarrow D_{3/2}$ Raman transition at a smaller detuning of $-298(1)$ MHz to resolve the AC-Stark shift at the lower intensity of the 408-nm laser. Results are shown in Fig. 6.10. The measured shift amounts to $-0{,}41(13)$ MHz/mW. Due to instabilities in the high voltage, the AC-stark shift of the $D_{3/2} \rightarrow P_{3/2}$ transition could not be extracted from the respective measurement series.

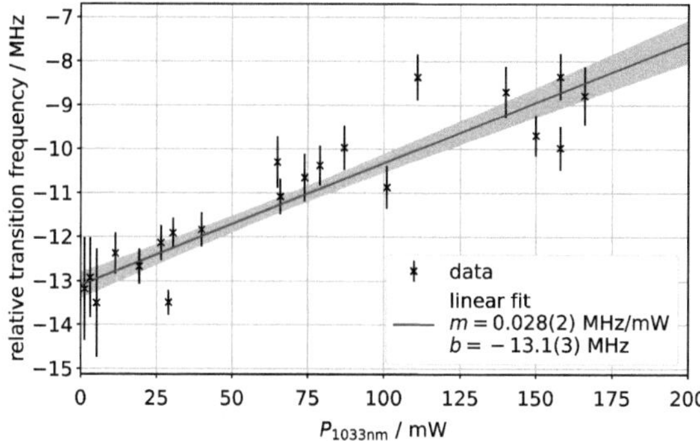

Fig. 6.9 Relative transition frequency of the $S_{1/2} \rightarrow D_{5/2}$ transition depending on the power P_{1033} of the 1033-nm laser, given in mW. The power of the 408-nm laser was set to $155(10)\,\mu W$. The measurements were performed in the optical pumping tube at a detuning of $-483(1)\,MHz$. The transition frequencies are given relative to $444\,779\,044\,MHz$. The black error bars indicate the experimental data and the fit uncertainties of the transition frequencies. To investigate the AC-Stark shift, a linear function $y = m \cdot P_{1033} + b$ was fitted. The fit result is shown with its confidence interval in red. The corresponding fit values are listed in the legend of the plot

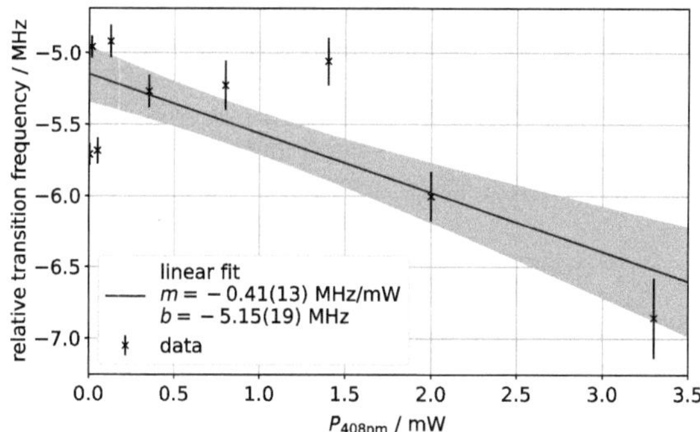

Fig. 6.10 Relative $S_{1/2} \rightarrow D_{3/2}$ transition frequency depending on the power P_{408} of the 408-nm laser, given in mW. The measurements were performed in the optical pumping tube at a detuning of $-298(1)\,MHz$ and at a laser power of the 1004-nm laser of $355(3)\,mW$. The black error bars mark the measured transition frequencies, given relative to $436\,373\,693\,MHz$. A linear function $y = m \cdot P_{1004} + b$ was fitted and is shown with its confidence interval. Due to technical difficulties with the high voltage stabilization, only a fraction of the data taken could be used for the analysis

A third set of measurements was performed using the double Raman scheme to investigate the AC-stark shift in the $S_{1/2} \to D_{3/2}$ as well as the $D_{5/2} \to D_{3/2}$ Raman transitions. The same settings as for the measurements presented in Sect. 5.2 were used. In Fig. 6.11 the results for the dependence of the $S_{1/2} \to D_{3/2}$ transition frequency from the power of the 1004-nm Matisse laser at a detuning of 1091(1) MHz are presented. From the slope of the linear fit, a shift of $-0,0031(1)$ MHz/mW is deduced.

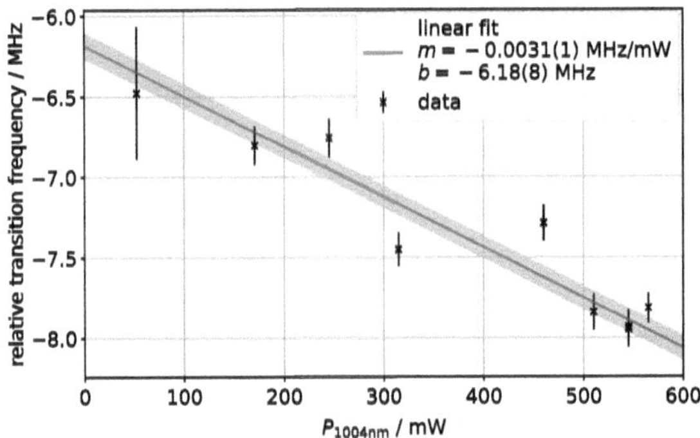

Fig. 6.11 Measured $S_{1/2} \to D_{3/2}$ transition frequency, depending on the power P_{1004} of the 1004-nm laser, given in mW. The measurements were performed in the second pumping region at a detuning of 1091(1) MHz and at a laser power of 2,5(1) mW for the 408-nm laser. The transition frequencies are given relative to 436 373 693 MHz. A linear function $y = m \cdot P_{1004} + b$ was fitted and is shown in red. The shaded area indicates the confidence interval of the fit

Using the fiber-coupled diode laser imposes limits on both laser power (max. 20 mW) and frequency stability (see Sect. 6.4.1). Thus, no power-dependent shift induced by the 1033-nm laser, nor the shift induced by the 1004-nm laser in the $D_{5/2} \to D_{3/2}$ transition could be quantified.

A direct comparison of the AC-Stark shifts to literature values is not possible, since, due to the spatial intensity distribution of the laser beams, different ions interact at different laser intensities, and thus only an effective average AC-Stark shift is measured. The ratio of the measured slopes can be compared to the ratio of the squares of the dipole moments calculated using 2.9 and the values listed in Tab. 2.1 and 2.2. To account for the different detunings Δ and laser beam diameters d at which the measurements were performed, the shifts are scaled with $\Delta \cdot d^2$. For the 408-nm $S_{1/2} \to P_{3/2}$ and the 1033-nm $D_{5/2} \to P_{3/2}$ transitions the predicted dipole ratio is 3.0 and the ratio of the rescaled slopes is 2.8(9), yielding excellent agreement, despite the large uncertainty in the slope of the 408-nm shift.

For the 1004-nm $D_{3/2} \rightarrow P_{3/2}$ transition and the 1033-nm $D_{5/2} \rightarrow P_{3/2}$ transitions a ratio of 0.16 is expected, and a ratio 0.20(3) was measured, also yielding good agreement.

Absolute transition frequency measurements are corrected by scaling the measured shifts with the diameter of the laser beams and the detuning, as well as the laser power, at which the measurements were performed:

$$\delta f_{\mathrm{AC},1} = m_{\mathrm{AC}} \cdot \frac{\Delta_0}{\Delta} \frac{d_0^2}{d^2} \cdot P. \tag{6.8}$$

Δ_0 and d_0^2 are the detuning and laser beam diameter at which the slope m_{AC} was measured, and Δ, d, and P are the detuning, laser beam diameter, and laser power at which the measurement to be corrected was performed.

The expression given Eq. 2.19 is a first-order approximation of the full AC-Stark shift for $\Omega_{ii} \ll \Delta$, that can be obtained using the dressed-state approach [24]. Measurements of the $D_{5/2} \rightarrow D_{3/2}$ were conducted at a detuning of only -175 MHz and at high laser powers. Thus, the second-order AC-stark shift has to be accounted for. A Taylor expansion up to the second order for $\Omega_{ii} << \Delta$ yields

$$\Omega_n^{AC} = \frac{|\Omega_{ii}|^2}{4\Delta} - \frac{|\Omega_{ii}|^4}{16\Delta^3}. \tag{6.9}$$

This can be expressed in terms of the first-order approximation $\Omega_n^{AC,1} = \frac{|\Omega_{ii}|^2}{4\Delta}$ as

$$\Omega_n^{AC} = \Omega_n^{AC,1} \left(1 - \left| \frac{\Omega_n^{AC,1}}{\Delta} \right| \right). \tag{6.10}$$

The correction of the AC-Stark shift up to the second order is hence given by $\delta f_{\mathrm{AC}} = \delta f_{\mathrm{AC},1} \left(1 - \left| \frac{\delta f_{\mathrm{AC},1}}{\Delta} \right| \right)$. For the $D_{5/2} \rightarrow D_{3/2}$ transition measurements, the second order contribution is of several $100\,\mathrm{kHz}$. For all other transitions, the second order was also included in the analysis but is negligible compared to systematic uncertainties discussed in the next paragraph.

Since the laser power was not measured after every measurement, a daily average value was used for the power P of each laser. The uncertainties listed in Tab. 6.5 were considered to introduce a systematic uncertainty for the AC-Stark shift. From the listed uncertainties, a total relative uncertainty is calculated. This uncertainty is then applied to the transition frequency measurements depending on the laser power used. This accounts for the increase of the AC-Stark shift and the connected uncertainty with increasing laser power.

Table 6.5 Uncertainties included in the estimate of the systematic uncertainty of the AC-Stark shift. Relative uncertainties are listed since scaling depending on the laser power is required

uncertainty	value
fit uncertainty in m_{AC}	see Fig. 6.9, 6.10, 6.11
laser beam diameter	$\frac{\delta d}{d} \leq 0.1$
laser power	drifts by typically 10%
detuning Δ	$\frac{\delta \Delta}{\Delta} < 1\%$

The voltage calibration described in Sect. 6.4.4 is—in theory—also susceptible to the AC-Stark shift. A series of consecutive voltage calibrations were performed at $P_{408} = 55\,\mu W$ and $P_{408} = 3\,mW$, resulting in an offset of $1{,}25(2)\,V$ and $1{,}24(1)\,V$. This does not contradict the AC-stark shift measured in Fig. 6.10. While the AC-Stark shift increases for $|\,\Delta\,| \rightarrow 0$, it also changes sign at $\Delta = 0$. At $\Delta = 0$, the AC-Stark splitting in the dressed-states approach is symmetric. As this splitting can not be resolved, it is expected that the peak is broadened but not shifted.

For the first time, this demonstrates that a Raman transition has been driven in a fast ion beam using two separate lasers, which confirms the feasibility of exploiting Raman transitions at COALA.

The derived analytical description of Raman transitions as two-photon Rabi oscillations is consistent with numerical solutions of the Liouville equation of the full Hamiltonian within the made approximations, and allows studying the impact of ion energy distributions as well as spatial intensity distributions on two-photon Rabi oscillations and the experimentally expected line shape. These simulations correctly predict the "collapse" of Rabi oscillations, which has been experimentally confirmed within this work.

It has also been shown, that Raman transitions can reach similar efficiencies in population transfer as optical pumping. This is of particular interest in elements, in which optical pumping is disfavored due to low branching ratios. A possible candidate for exploiting this would be thulium. Exploiting state selective charge exchange will give further insight into the efficiency of the population transfer.

While the sub-200kHz precision of recent high-precision collinear/anticollinear measurements on dipole transitions at COALA [33, 38, 49] has not been reached yet, based on the measurements performed in this work, this could be achieved by investigating the AC-Stark shift with a frequency comb, and by performing collinear/anticollinear measurements at a higher detuning of $1 - 2\,\text{GHz}$. Nevertheless, the measurements performed on the $S_{1/2} \rightarrow D_{5/2}$ clock transition have demonstrated, that Raman transitions can be used to perform high-precision collinear laser spectroscopy in dipole-forbidden transitions and yield excellent agreement with literature values.

Furthermore, a new scheme to perform Doppler-free Raman transitions has been developed and was used for the first direct $D_{5/2} \rightarrow D_{3/2}$ transition frequency

J. Spahn, *Pioneering Raman Transition Techniques in Collinear Laser Spectroscopy for High-Precision Experiments*, BestMasters,
https://doi.org/10.1007/978-3-658-50605-6_7

measurement in $^{88}\text{Sr}^+$. The double Raman scheme allowed to investigate the entire ring closure between the $S_{1/2}$, $D_{3/2}$, and $D_{5/2}$ states in $^{88}\text{Sr}^+$ within a single measurement. The obtained frequencies were constrained using this ring closure and the collinear/anticollinear measurements in the $S_{1/2} \rightarrow D_{5/2}$ transition, resulting in values of $8\,505\,350.8(7)$ MHz, $436\,373\,693.4(6)$ MHz, and $444\,779\,044.1(4)$ MHz for the $D_{5/2} \rightarrow D_{3/2}$, $S_{1/2} \rightarrow D_{3/2}$, and $S_{1/2} \rightarrow D_{5/2}$ transitions, respectively.

Currently, linewidths measured in this scheme are limited to about 3.5 MHz by temporal and spatial inhomogeneities in the potential of the interaction regions and by time-of-flight broadening. In the currently ongoing redesign of the post-acceleration region at COALA, that will be used for future high-precision high-voltage measurements, these aspects will be considered and suitable pumping sections will be integrated. This would allow performing spectroscopy with sub-megahertz linewidths, which will allow high precision collinear laser spectroscopy studies and can be used as velocity filter to improve the resolution of the high-voltage studies [14].

Bibliography

1. K-R Anton et al. "Collinear laser spectroscopy on fast atomic beams". In: *Physical Review Letters* **40** (1978), p. 642. https://doi.org/10.1103/PhysRevLett.40.642.
2. XF Yang et al. "Laser spectroscopy for the study of exotic nuclei". In: *progress In Particle And Nuclear Physics* **129** (2023), p. 104005. https://doi.org/10.1016/j.ppnp.2022.104005.
3. *cls On Radioactive Nuclei—Overview*. Accessed: 2024-04-04. URL:www.ikp.tu-darmstadt.de/_laser_nuclear_chart.
4. W Nörtershäuser. "Schnelle Ionen im Laserlicht". In: *Physik Journal* **17** (2018), p. 33 ff.
5. Alex C Mueller et al. "Spins, moments and charge radii of barium isotopes in the range $^{122--146}$ Ba determined by collinear fast-beam laser spectroscopy". In: *Nuclear Physics A* **403** (1983), p. 234–262. https://doi.org/10.1016/0375-9474(83)90226-9.
6. Klaus Blaum, Jens Dilling, and Wilfried Nörtershäuser. "Precision atomic physics techniques for nuclear physics with radioactive beams". In: *Physica Scripta* **2013** (2013), p. 014017. https://doi.org/10.1088/0031-8949/2013/T152/014017.
7. Á Koszorús et al. "Charge radii of exotic potassium isotopes challenge nuclear theory and the magic character of N= 32". In: *Nature Physics* **17** (2021), pp. 439–443. https://doi.org/10.1038/s41567-020-01136-5.
8. Kristian König et al. "Surprising charge-radius kink in the Sc isotopes at N= 20". In: *Physical Review Letters* **131** (2023), p. 102501. https://doi.org/10.1103/PhysRevLett.131.102501.
9. Adolf Smekal. "Zur Quantentheorie der Dispersion". In: *Naturwissenschaften* **11** (1923), pp. 873–875. https://doi.org/10.1007/BF01331666.
10. C Raman. *Nobel Lectures: Physics 1922–1941*. 1965.
11. David E Bugay. "Characterization of the solid-state: spectroscopic techniques". In: *Advanced Drug Delivery Reviews* **48** (2001), pp. 43–65. https://doi.org/10.1016/S0169-409X(01)00101-6.
12. Xiaoshan Zheng et al. "Raman imaging from microscopy to nanoscopy, and to macroscopy". In: *Small* **11** (2015), pp. 3395–3406. https://doi.org/10.1002/smll.201403804.
13. J Reichel et al. "Subrecoil Raman cooling of cesium atoms". In: *Europhysics Letters* **28** (1994), p. 477. https://doi.org/10.1209/0295-5075/28/7/004.

14. Antje Neumann, R Walser, and W Nörtersh user. "Raman velocity filter as a tool for collinear laser spectroscopy". In: *Physical Review A* **101** (2020), p. 052512. https://doi.org/10.1103/PhysRevA.101.052512.

15. K König et al. "First high-voltage measurements using Ca$^+$ ions at the ALIVE experiment". In: *Hyperfine Interactions* **238** (2017), pp. 1–12. https://doi.org/10.1007/s10751-016-1392-4.

16. TP Dinneen et al. "Stimulated Raman measurements of the hyperfine structure in Yi I". In: *Physical Review A* **43** (1991), p. 4824. https://doi.org/10.1103/PhysRevA.43.4824.

17. Philipp Bollinger. "Kollineare Raman-Spektroskopie an ^{40}Ca$^+$ und Optimierung der Hochspannungsmessung durch systematische Minimierung der Ionenstrahldivergenz". TU Darmstadt, 2017.

18. L Vermeeren et al. "Ultrasensitive radioactive detection of collinear-laser optical pumping: measurement of the nuclear charge radius of ^{50}Ca". In: *physical Review Letters* **68**.11 (1992), p. 1679. https://doi.org/10.1103/PhysRevLett.68.1679.

19. RF Garcia Ruiz et al. "Development of a sensitive setup for laser spectroscopy studies of very exotic calcium isotopes". In: *Journal of Physics G: Nuclear and Particle Physics* **44** (2017), p. 044003. https://doi.org/10.1088/1361-6471/aa5a24.

20. TE Cocolios et al. "The collinear resonance ionization spectroscopy (CRIS) experimental setup at CERN-ISOLDE". In: *Nuclear Instruments and Methods in Physics Research Section B: Beam Interactions with Materials and Atoms* **317** (2013), pp. 565–569. https://doi.org/10.1016/j.nimb.2013.05.088.

21. Wolfgang Demtröder. Laserspektroskopie: *Grundlagen und Techniken*. Springer, 2007.

22. Kristian Lars K nig. "Laser-Based High-Voltage Metrology with ppm Accuracy". en. PhD thesis. Darmstadt: Technische Universität Darmstadt, 2019. https://doi.org/10.26083/tuprints-00008401.

23. J Krämer et al. "High-voltage measurements on the 5 ppm relative uncertainty level with collinear laser spectroscopy". In: *Metrologia* **55**.2 (2018), p. 268. https://doi.org/10.1088/1681-7575/aaabe0.

24. Alexander J Dunning. *Coherent Atomic Manipulation and Cooling: Interferometric Laser Cooling and Composite Pulses for Atom Interferometry*. Springer, 2015. https://doi.org/10.1007/978-3-319-21738-3.

25. Bruce Shore. *The theory of coherent atomic excitation*. Taylor & Francis, 1991. https://doi.org/10.13182/FST91-A29400.

26. Daniel A Steck. "Sodium D line data". In: *Report, Los Alamos National Laboratory, Los Alamos* **124** (2000).

27. Andreas Bauch. "Caesium atomic clocks: function, performance and applications". In: *Measurement Science and Technology* **14** (2003), p. 1159. https://doi.org/10.1088/0957-0233/14/8/301.

28. Bruce W. Shore. *Manipulating Quantum Structures Using Laser Pulses*. Cambridge University Press, 2011.

29. Stephen M. Barnett and Paul M. Radmore. "Dressed States". In: *Methods in Theoretical Quantum Optics*. Oxford University Press, Nov. 2002. https://doi.org/10.1093/acprof:oso/9780198563617.003.0006.

30. Roger C Brown et al. "Quantum interference and light polarization effects in unresolvable atomic lines: Application to a precise measurement of the ^{6,7}Li D2 lines". In: *Physical Review A* **87** (2013), p. 032504.

31. K König et al. "A new collinear apparatus for laser spectroscopy and applied science (COALA)". In: *Review of Scientific Instruments* **91** (2020), p. 081301. https://doi.org/10.1063/5.0010903.

32. Julien Spahn. "Untergrundbereinigte kollineare Sättigungsspektroskopie an $^{12}C^{4+}$". TU Darmstadt, 2022.

33. Julian Palmes. "Collinear Laser Spectroscopy of singly charged strontium isotopes". TU Darmstadt, 2023.

34. Juris Meija et al. "Isotopic compositions of the elements 2013 (IUPAC Technical Report)". In: *Pure and Applied Chemistry* **88** (2016), pp. 293–306. https://doi.org/10.1515/pac-2015-0503.

35. Meng Wang et al. "The AME 2020 atomic mass evaluation (II). Tables, graphs and references". In: *Chinese Physics C* **45** (2021), p. 030003.

36. Jagdish K Tuli et al. *Nuclear wallet cards.* Brookhaven National Laboratory Upton, NY, 1995.

37. A. Kramida et al. NIST Atomic Spectra Database (ver. 5.9), [Online]. Link: https://physics.nist.gov/asd [2022, April 05]. National Institute of Standards and Technology, Gaithersburg, MD. 2023. https://doi.org/10.18434/T4W30F.

38. Collinear laser spectroscopy of highly charged ions produced with an electron-beam ion source. "Collinear laser spectroscopy of highly charged ions produced with an electron-beam ion source". In: *Physical Review A* **108** (2023), p. 062809. https://doi.org/10.1103/PhysRevA.108.062809.

39. Patrick Matthias Müller. "Laserspectroscopic determination of the nuclear charge radius of ^{13}C". en. PhD thesis. Darmstadt: Technische Universität Darmstadt, 2024. https://doi.org/10.26083/tuprints-00026746.

40. Ingmar Metzler. "Realisierung des gepulsten Pump-Tast-Nachweises für das ALIVE-Experiment". TU Darmstadt, 2017.

41. Kristian König et al. "On the performance of wavelength meters: Part 2—frequency-comb based characterization for more accurate absolute wavelength determinations". In: *Applied Physics B* **126** (2020), p. 86. https://doi.org/10.1007/s00340-020-07433-4.

42. Phillip Imgram. "High-precision laser spectroscopy of helium-like carbon $^{12}C^{4+}$". PhD thesis. Technische Universität Darmstadt, 2023. https://doi.org/10.26083/tuprints-00023082.

43. Tim Ratajczyk. "Isotope shift measurements in Ti^+ transitions using laser-ablated and thermalized ions". PhD thesis. Technische Universität Darmstadt, 2021. https://doi.org/10.26083/tuprints-00020307.

44. CA Ryder et al. "Population distribution subsequent to charge exchange of 29.85 keV Ni+ on sodium vapor". In: *Spectrochimica Acta Part B: Atomic Spectroscopy* **113** (2015), pp. 16–21. https://doi.org/10.1088/1674-1137/abddaf.

45. Nadja Frömmgen. "Kollineare Laserspektroskopie an radioaktiven Praseodymionen und Cadmiumatomen". PhD thesis. Mainz U., 2013.

46. Sebastian Raeder. "Spurenanalyse von Aktiniden in der Umwelt mittels Resonanzionisationsmassenspektrometrie". PhD thesis. Mainz U., 2010.

47. Phillip Imgram. *Private communication.*

48. Simon Kaufmann. "Laser spectroscopy of nickel isotopes with a new data acquisition system at ISOLDE". PhD thesis. Technische Universität Darmstadt, 2019. https://doi.org/10.26083/tuprints-00009286.

49. Patrick M ller. "Collinear Laser Spectroscopy of Ca$^+$ Isotopes and Design of a Lens-Based Optical Detection Region". TU Darmstadt, 2019.

50. A Klose, K Minamisono, and PF Mantica. "Collinear laser spectroscopy on the ground state and an excited state in neutral ^{55}Mn". In: *Physical Review A—Atomic, Molecular, and Optical Physics* **88**.4 (2013), p. 042701. https://doi.org/10.1103/PhysRevA.88.042701.

51. Kristian König et al. "Transition frequencies and hyperfine structure in ^{113,115}In$^+$: Application of a liquid-metal ion source for collinear laser spectroscopy". In: *Physical Review A* **102**.4 (2020), p. 042802. https://doi.org/10.1103/PhysRevA.102.042802.

52. Pierre Dubé, John E Bernard, and Marina Gertsvolf. "Absolute frequency measurement of the ^{88}Sr$^+$ clock transition using a GPS link to the SI second". In: *Metrologia* **54**.3 (2017), p. 290.